International Conference on Interactive Design, Architecture and Engineering (ICIDAE 2021)

AA007297

Held online due to COVID-19

IOP Conference Series: Materials Science and Engineering
Volume 1148

Suzhou, China
17 - 19 March 2021

ISBN: 978-1-7138-3418-2
ISSN: 1757-8981

Printed from e-media with permission by:

Curran Associates, Inc.
57 Morehouse Lane
Red Hook, NY 12571

Some format issues inherent in the e-media version may also appear in this print version.

Institute of Physics
Dirac House, Temple Back
Bristol BS1 6BE UK

Phone: 44 1 17 929 7481
Fax: 44 1 17 920 0979

techtracking@iop.org

Additional copies of this publication are available from:

Curran Associates, Inc.
57 Morehouse Lane
Red Hook, NY 12571 USA
Phone: 845-758-0400
Fax: 845-758-2633
Email: curran@proceedings.com
Web: www.proceedings.com

TABLE OF CONTENTS

PREFACE

2021 2nd International Conference on Interactive Design, Architecture and Engineering (ICIDAE 2021) was successfully held at Suzhou, China during March 17-19, 2021. Due to the outbreak of COVID-19, the organizing committee will follow the guidelines of WHO (World Health Organization), and local government agencies on the current situation. The ICIDAE 2021 was hold virtually. All the keynote sessions and technical session of ICIDAE 2021 were successfully conducted through hybrid platform (Online and offline).

The objective of ICIDAE 2021 Conference is to provide a venue for researchers and practitioners in related fields from academia, industry and government to meet in a forum where the latest research results are presented and prospect for future developments on "interactive design, architecture". Professors from Portugal and Japan are invited to deliver keynote speeches regarding latest information in their respective expertise areas, where the speakers discussed various issues of architecture. There were one oral sessions. Based on the technical, quality and presentation, best papers were selected and awarded during each technical session.

This proceeding presents a selection from papers submitted to the conference from universities, research institutes and industries. All of the papers were subjected to peer-review by conference committee members and international reviewers. The papers selected depended on their quality and their relevancy to the conference. The volume tends to present to the readers the recent advances in the field of interactive design, architecture and various related areas.

We express our sincere gratitude to all people who contributed to the review process and did all their best to enhance the scientific merit and quality of the Proceedings. We are also grateful to the authors for their contributions, to the speakers and participants of the ICIDAE 2021. We hope to welcome them along with new participants to the next event ICIDAE 2022.

Prof. Nuno Dinis Corticos
Department of Technologies in Architecture, Urbanism and Design at Faculdade de Arquitectura, Universidade de Lisboa, Portugal
April 11, 2021

IOP Conf. Series: Materials Science and Engineering 1148 (2021) 011001 doi:10.1088/1757-899X/1148/1/011001

Conference Committee

IOP Conf. Series: Materials Science and Engineering 1148 (2021) 011001 doi:10.1088/1757-899X/1148/1/011001

IOP Conf. Series: Materials Science and Engineering 1148 (2021) 011001 doi:10.1088/1757-899X/1148/1/011001

Peer review declaration

All papers published in this volume of *IOP Conference Series: Materials Science and Engineering* have been peer reviewed through processes administered by the Editors. Reviews were conducted by expert referees to the professional and scientific standards expected of a proceedings journal published by IOP Publishing.

- **Type of peer review:** Double-blind (Conference reviewers and iThenticate system)
- **Conference submission management system:** Conference Management Toolkit (CMT)
- **Number of submissions received:** 15 full papers and 1 abstracts
- **Number of submissions sent for review:** 15 full papers
- **Number of submissions accepted:** 11 full papers
- **Acceptance Rate (Number of Submissions Accepted / Number of Submissions Received X 100):** 73%
- **Average number of reviews per paper:** 3
- **Total number of reviewers involved:** 78
- **Any additional info on review process:** The papers were evaluated based on the technical criteria, quality criteria and presentation criteria. The referees argued detailed their reviews and provided an overall evaluation (strong accept, accept, weak accept, weak reject, reject, strong reject), specifying their confidence (expert, high, medium, low, none) regarding qualifications to review the allocated paper.

The criteria the reviewers used to assess the papers as following:

- **1. The Topic's Conformity, The Coverage of the Topic**
- **2. Innovation, Integrality, The "literary" presentation, The technical depth of content.**
- **3. The English grammar.**
- **4. Originality, Motivation, Repetition, Length of quality.**
- **5. Scientific merit, Clarity of expression, Referencing.**

- **Contact person for queries:** Ms. Alice Liu; alice@isaeri.org

The use of artificial intelligence (AI) and Big-Data to improve energy consumption in existing buildings

Mamdooh Alwetaishi[1a] **and Amal Shamseldin**[1, 2b]
[1]Department of Civil Engineering, College of Engineering, Taif University, P.O. Box 11099, Taif 21944, Saudi Arabia
[2]Archtectural Engineering, Department, College of Engineering, Ain Shams University, Cairo, 11566, Egypt
[a]m.alwetaishi@tu.edu.sa; [b]amal.kamal@eng.asu.edu.eg

Abstract. The use of Artificial Intelligence AI and Big data can contribute to improve energy performance in buildings. There are so many data are available in different institutions within a country such as Ministry of housing and electricity companies. The aim of this research is to propose a link between all these institutions with the aid of field measurement and occupants' thermal sensation. The work used various methods to carry out this research including field measurement, computer modelling and survey. It was found that use of AI in existing buildings can improve energy performance significantly. The current research focused on optimization of WWR (Window to Wall Ratio) with orientation. WWR for western zones should not exceed 15% in extreme hot and humid climate and it can reduce energy consumption by 20%.

1. Introduction

Human performance is effected substantially by thermal comfort on the level of physical and mental [1]–[3]. One of the major reasons for this is that we are as human spends more than 90% of our times in doors. There are four major aspects contribute to the total health and productivity of people which are all linked to indoor environmental quality (IEQ): thermal comfort, visual comfort indoor air quality and acoustics (quality of sound and noise). Based on an abundant research publication and previous studies, there is no doubt that human thermal comfort is the most important aspect among all of them [4], [5]. Buildings consume roughly 40% of total energy globally to achieve thermal comfort [6]. As a result, thermal comfort is very linked to Artificial Intelligence AI to achieve comfort level with less amount of energy consumed [4], [7], [8]. In addition to that it can aid to increase productivity in living spaces indoor environment [9].

Figure 1. Location of residential zone in Jeddah

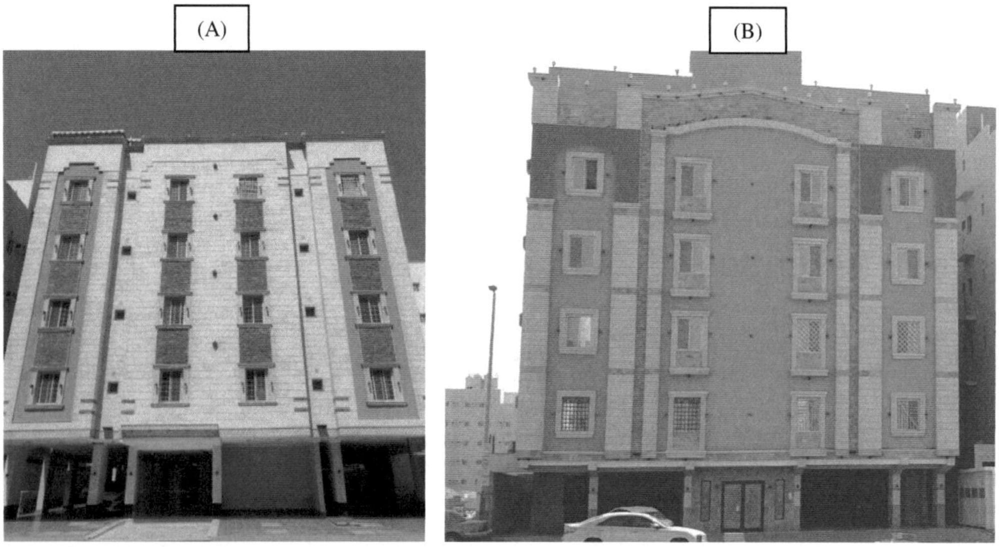

Figure 2. Residential blocks with different orientation (A) is west and (B) is north, it is clear that all facades are similar in terms of design and openings size

There are various methods used to investigate AI in buildings such as genetic programming, neural networks, fuzzy logic, heuristic search, computer software and many others. The use of AI is not only restricted and associated with thermal comfort, it is also linked to safety in buildings [10]. In a study conducted by [11] highlighted that AI to indoor thermal comfort. In addition to that, the work suggested that there are additional variables that can be further investigated such as lighting control especially when related to indoor luminance. The work of Orosa [12] presented a novel method based on Neural networks, it shows that a possible 20% of energy consumption in HVAC in buildings is possible utilising this system in order to achieve more comfortable indoor environment in summer. Another work conducted in tropical climate of Singapore [13] it showed that using neural network AI, it was found that 25°C is an ideal indoor temperature with minimum use in energy consumption. Marinakis [14] presented an investigation of use of internet of things for energy efficient buildings, it aids to improve reliable and effective policymaking in data-driven architecture. Paper also reported the importance of adaptation of big data such as Internet of things (IoT), distributed technology (DLT) and Artificial Intelligence (AI). BIM (Building Information Modelling) was found to be compatible with AI. BIM from mid of 1980s used to be part of AI. However, in the past five years start-up companies named it as Construction Tech [15].

2. Method and materials used

2.1. Location and case studies

The study was conducted in a residential building blocks located in the city of Jeddah (Figure 1). The climatic of the city is hot and humid with very hot temperatures in summer and warm in winter. The study selected two building blocks (Figure 2) which are consisting of five storeys, they comparable in size, building materials and differ in orientation with west and north orientations as identified in the figure.

2.2. Equipment used for monitoring and survey

This study is carried out using data-loggers for site monitoring for indoor air temperature and relative humidity. The field measurement and survey took place in the city of Jeddah on Jun 2020; the hottest months were targeted to obtain extreme results and feedback. In addition to that, a survey for selected families was conducted to obtain thermal comfort sensation in air-conditioning operation mode. There were about 18 families were asked on site to observe thermal comfort level for two different apartments with west and north orientations.

2.3. Computer modeling

The computer modeling software TAS EDSL was used for energy evaluation performance. It is also used to calculate amount of solar energy in two different facing properties.

3. Discussion and results

Results of data-loggers (Figure 3) indicate that there is a clear influence of orientation in residential buildings in hot regions. The monitoring took place for seven days of duration with a slot of an empty property to have the AC system switch off to be in the free running mode. It can be noticed that in the north facing apartment, the indoor temperature reached only 33.1°C while in west direction reached as high as 35.5°C. This is due to direct exposure to solar radiation on the west orientation which is considered as one of the worst in hot regions [16], [17]. Humidity also affected by raise of indoor temperature in reverse relation, it will impact the state of human thermal comfort and hence, amount of energy used in buildings. This can be seen in Figure 4 where total amount of energy used in the selected apartments are presented. It was found that there is an increase in west apartment compared to north by 20%. It has to be mentioned that both area and family size of two sleeted apartments were comparable.

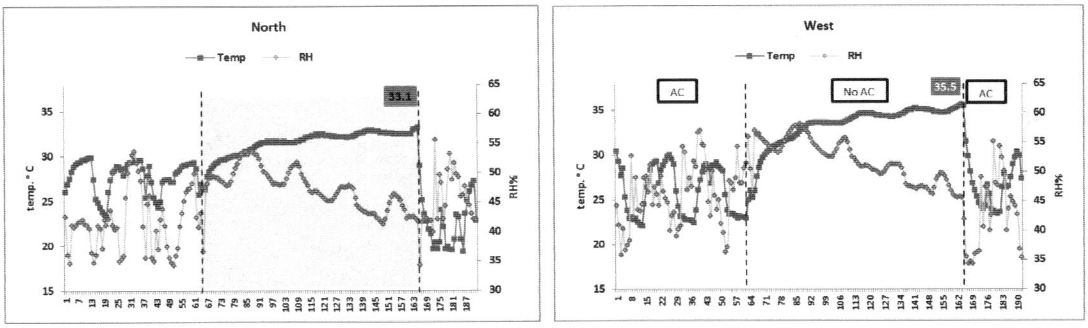

Figure 3. Data-loggers monitoring in north and south facing apartments

Thermal comfort votes PMV (Predicted Mean Vote) also investigated as it can be noted in Figure 5. The feedback of occupants was taken during the operation of air-conditioning system. However, the west facing apartment influenced to thermal comfort due to heat gain from external envelope in the west facades from external walls as well as glazing system which is responsible for considerable amount of energy consumed. As a result, it is recommended that WWR in western facing rooms should be minimized to only allow adequate amount of daylight. In the work of Alwetaishi [17] WWR should not exceed 20%, in fact, this study is conducted in a hot region but it was a high altitude mountains, in a more heated outdoor condition like the case in the city of Jeddah, WWR should not exceed 15% which is quite similar to the recommendations of a study conducted in the city of Riyadh [16]. Figure 6 shows the amount of solar of both north and west. Size of windows in this region is very crucial and should be determined precisely.

Proposed system (Figure 7) present a new and innovative approach to control and provide a dynamic solution for architectural early stage wrong decision such as window to wall ratio. This can also be implemented on larger number of architectural elements such as thermal insulation, shading devices and thermal leakage. The system relies on AI and use of big data provided in the form of amount of electricity used by each property in a residential area. This can be formed by Ministry of housing which can group each area based on size of property and family members. MH then will identify each family member who is above average level of energy use to be reported to Sustainable Building Section (SBS). They have a lot of experts who can propose recommendations to improve energy performance of each property based on thermal comfort survey feedback conducted remotely. The system can contribute significantly on improving the energy performance as well as thermal comfort level for occupants.

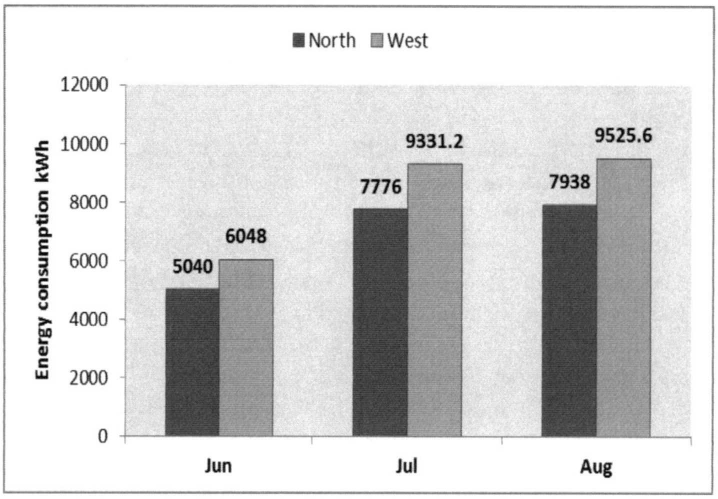

Figure 4. Energy consumption in the two facing apartments, kWh

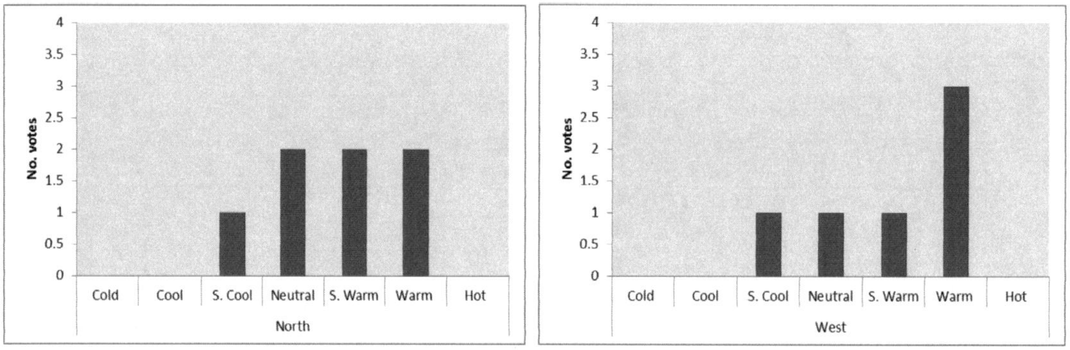

Figure 5. PMV of occupants based with operative system

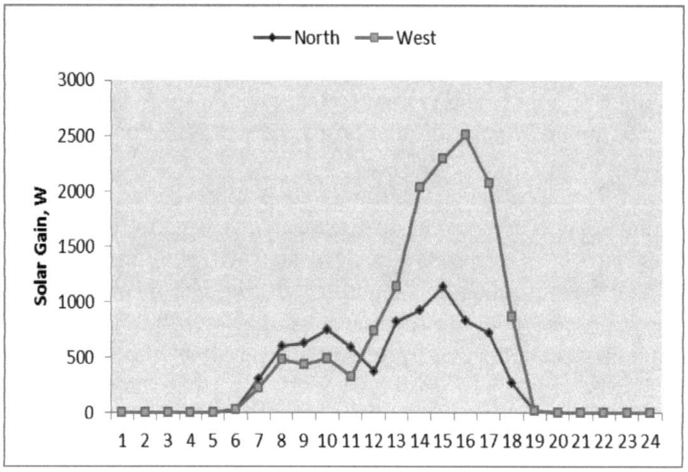

Figure 6. Solar heat gain in north and west facing zones in summer

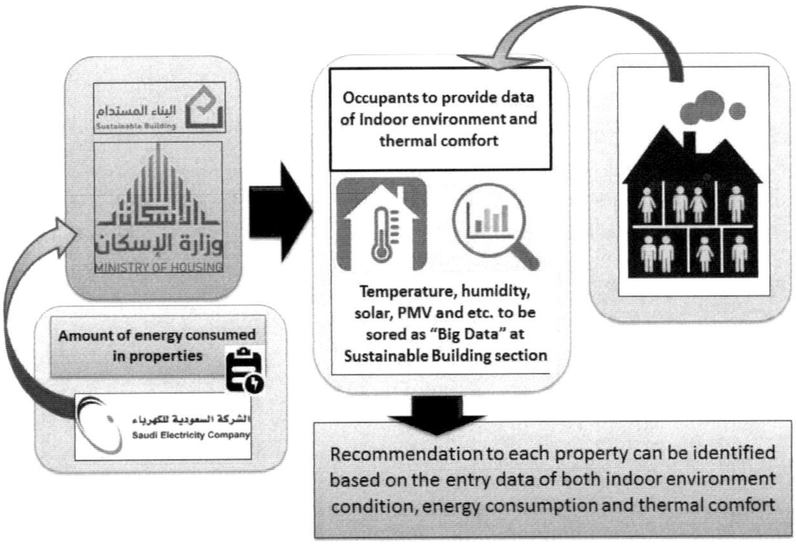

Figure 7. Proposed system of AI system based on indoor environment, comfort and energy consumption

4. Conclusion

This study investigated the potential to apply AI to improve energy performance as well as thermal comfort using new system utilizing "big data" and AI. There are several institutions which are involved starting from Ministry of housing which has the responsibility to obtain energy consumption for all properties and submit them to sustainable building section. The SBS can study the data based on micro-locations and size of property as well as family. Field visit might be included to observe architectural and conduction condition. In the current investigated properties, west and north facing apartments were presented. It was found that amount of energy consumed in the west facing apartment was higher by 20%. It is suggested that WWR of western facing zones should not be greater than 15% in extreme hot regions such as in Jeddah where outdoor temperatures can get higher than 40°C in daytime during summer.

5. List of References

[1] Banerjee R. Importance of Net Zero Energy Building. *Int J Innov Res Adv Eng* 2015; 2: 141–145.

[2] Pisello AL, Pignatta G, Castaldo VL, et al. The impact of local microclimate boundary conditions on building energy performance. *Sustain* 2015; 7: 9207–9230.

[3] Alwetaishi M, Balabel A, Abdelhafiz A, et al. User thermal comfort in historic buildings: Evaluation of the potential of thermal mass, orientation, evaporative cooling and ventilation. *Sustain* 2020; 12: 1–24.

[4] Ghahramani A, Galicia P, Lehrer D, et al. Artificial Intelligence for Efficient Thermal Comfort Systems: Requirements, Current Applications and Future Directions. *Front Built Environ*; 6. Epub ahead of print 2020. DOI: 10.3389/fbuil.2020.00049.

[5] Oancea C, Caluianu S. Analysis of non-residential buildings in Romania from the labour productivity and intelligent buildings concept point of view. *Intell Build Int* 2012; 4: 216–227.

[6] Ngarambe J, Yun GY, Santamouris M. The use of artificial intelligence (AI) methods in the prediction of thermal comfort in buildings: energy implications of AI-based thermal comfort controls. *Energy Build* 2020; 211: 109807.

[7] Li N, Yu W, Li B. Assessing adaptive thermal comfort using artificial neural networks in naturally-ventilated buildings. *Int J Vent* 2012; 11: 205–217.

[8] Zhao Y, Genovese PV, Li Z. Intelligent thermal comfort controlling system for buildings based on IoT and AI. *Futur Internet*; 12. Epub ahead of print 2020. DOI: 10.3390/fi12020030.

[9] Dounis AI. Artificial intelligence for energy conservation in buildings. *Adv Build Energy Res* 2010; 4: 267–299.
[10] Panchalingam R, Chan KC. A state-of-the-art review on artificial intelligence for Smart Buildings. *Intell Build Int* 2019; 1–24.
[11] Kolokotsa D. Artificial intelligence in buildings: A review of the application of fuzzy logic. *Adv Build Energy Res* 2007; 1: 29–54.
[12] Orosa JA, Vergara D, Costa ÁM, et al. A novel method based on neural networks for designing internal coverings in buildings: Energy saving and thermal comfort. *Appl Sci*; 9. Epub ahead of print 2019. DOI: 10.3390/app9102140.
[13] Chaudhuri T, Soh YC, Li H, et al. A feedforward neural network based indoor-climate control framework for thermal comfort and energy saving in buildings. *Appl Energy* 2019; 248: 44–53.
[14] Marinakis V. Big data for energy management and energy-efficient buildings. *Energies*; 13. Epub ahead of print 2020. DOI: 10.3390/en13071555.
[15] Sacks R, Girolami M, Brilakis I. Building Information Modelling, Artificial Intelligence and Construction Tech. *Dev Built Environ* 2020; 100011.
[16] Alwetaishi M. Impact of glazing to wall ratio in various climatic regions: A case study. *J King Saud Univ - Eng Sci* 2019; 31: 6–18.
[17] Alwetaishi M, Taki A. Investigation into energy performance of a school building in a hot climate: Optimum of window-to-wall ratio. *Indoor Built Environ* 2020; 29: 24–39.

Designing a facade by biomimicry science to effectively control natural light in buildings (Glare analysis)

Sukhum Sankaewthong*[1a], Teerayut Horanont[1], Kazunori miyata[2] and Jessada Karnjana[3]

[1]School of Information, Computer and Communication Technology, Sirindhorn International Institute of Technology, Thammasat University, Thailand
[2] School of knowledge science, Japan Advanced Institute of Science and Technology, Ishikawa, Japan
[3] National Electronics and Computer Technology Centre (NECTEC), Thailand
[a]Sukhumphed@gmail.com

Abstract. The increasingly popular design trend of glazed facades using daylight in buildings has made it essential for innovations to eliminate the unnecessary intrusion of sunlight in spaces. This study aims to create a kinetic façade pattern by referring DNA structure and photosynthetic behaviour to mimic biomimicry science characteristics in *Wallacei* evolutionary software for generating possible patterns. Daylight glare thresholds were determined as an essential factor for user productive work. Comparison of three-building envelope potential was made for preparing spaces (zones A, B) in Bangkok condition; without a façade, with a static facade, and with a kinetic façade. *DIVA* software was used to analyse glare in terms of daylight glare probability (DGP). First, DGP (without façade) for zones A and B were 100% and 55%, or intolerable glare. Second, DGP (static façade) for zones A and B were 59%, 30%; zone A was intolerable, and B imperceptible. Third, DGP (kinetic façade) for zones A and B were 28% (imperceptible glare). Therefore, a kinetic façade has a high potential for protecting against unsuitable glare. These findings may serve as preliminary evidence for understanding kinetic façade potential for self-adjustment by light intensity to improve quality of life for occupant use of spaces.

1. Introduction

Buildings are now designed with either exaggerated proportions of transparent materials or large open spaces within a building envelope [1], [2]. For instance, in the images of Suvarnabhumi International Airport (Figure 1) and a working space (Figure 2), there are various issues that can occur from excessive natural light, one of which is "intolerable glare". Glare directly affects human activities that use spaces such as these as working spaces, creation spaces, and relaxing spaces, yet this is particularly so in working spaces since people must focus their eyesight [3]. Natural light is crucial since it can make work more effective. In 2011, a poll was conducted among 135 architects, lighting designers, and consultants, which found that over 80 percent of participants deemed natural light to be either an important or extremely important design factor (Figure 3). Glare factor from daylight was simulated for participants in various fields, while the simulations were integrated with luminance contrast measurements and direct sunlight identification in the users' proximity in a workspace [4]. Natural light is related to human health, especially eyestrain which is a common condition that occurs when the eyes get tired from intense use, such as staring at computer screens and other digital devices, reading books, or working. Almost all such problems occur as a consequence of unsuitable light, including natural and artificial light that can result in eye discomfort [5]. These issues arise since open areas of wall are unsuitable for a controlled climate. Therefore, the use of spaces are inefficient to ensure that natural light remains suitable for human activities in such spaces. This issue prompted the researcher's interest to suitably control the atmosphere in buildings, specifically the glare factor which affects human comfort, by using the façade to control this problem.

Figure 1. Suvarnabhumi International Airport, Thailand has an exaggerated proportion of transparent materials.

Figure 2. Over glazed façades lead to spaces with excessive glare or overheating. Occupants have no choice but to install personal shade equipment.

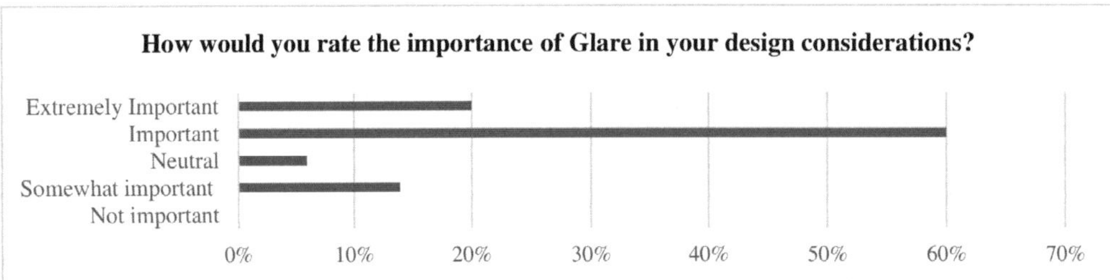

Figure 3. Poll of 135 architects, lighting designers, and consultants on the importance of glare *(Mogri 2011).*

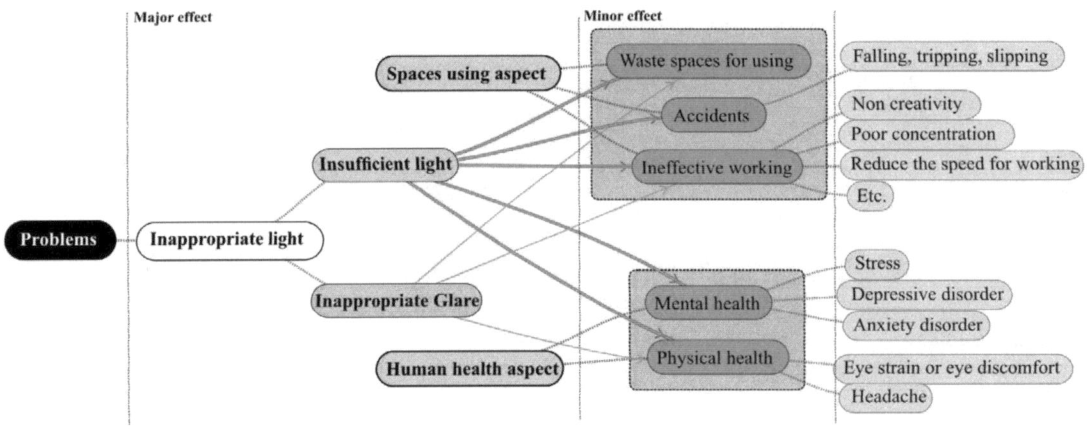

Figure 4. The problem uses of unsuitable glazed panels in the building envelope.

Figure 4 presents the significant problem posed by inappropriate light. Figure 4 separates this into two issues, insufficient light and inappropriate glare. First, an insufficient light issue affects both the use of spaces and human health aspects. The muse of space aspect is then divided into three issues: (1) unusable space; (2) accidents such as falling, tripping, and slipping; and (3) ineffective working including non-creativity, poor concentration, and reduced working speed. The human health aspect is then separated into two further issues: (1) mental health such as stress, depressive disorder, and anxiety disorder; and (2) physical health, including eye strain or eye discomfort, and headache. Second, problematic glare issues also affect both aspects [6]. Nevertheless, some minor issues are not included, namely the accidental and mental health aspects.

2. Study purpose

The research studies and mimics natural form and movement. This study focuses on DNA structure and the phototropism phenomenon to find a distinctive point by interpreting its physical appearance. Kinetic façade technology is developed to effectively controlled a suitable interior atmosphere that has natural light in indoor spaces in order to encourage work efficiency and human health. The purposes of this study are as follows:

2.1. To research potential façade patterns (form-finding) that effectively provides a suitable interior environment that has natural light, with a focus on the glare factor.

2.2. To study natural movement (biomimicry) to interpret a suitable façade movement.

2.3. To analyse the effectiveness of the kinetic façade in terms of glare.

3. Research methodology

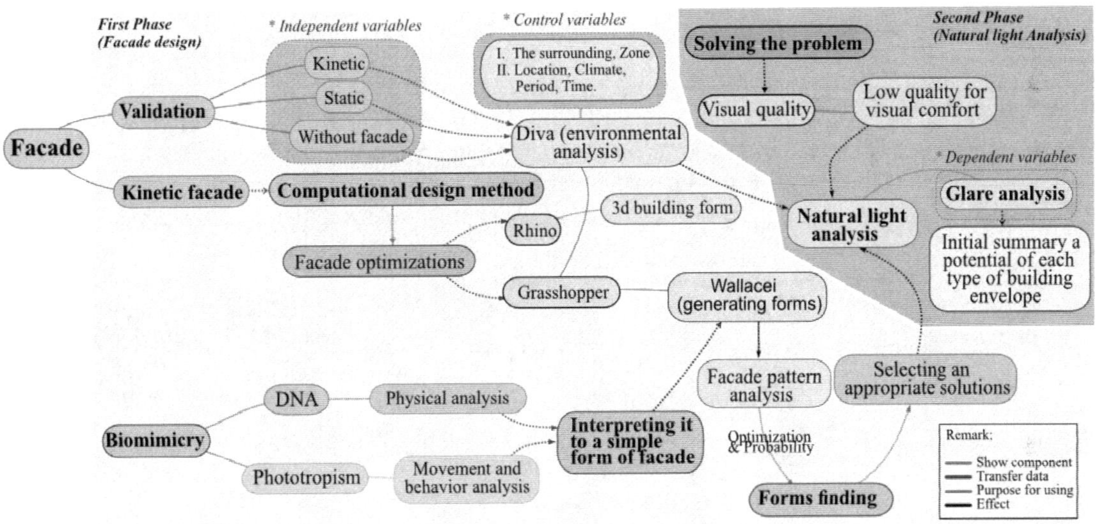

Figure 5. The research methodology.

Figure 5 presents the overall research methodology to determine how a façade directly affects the glare factor. The method is divided into two phases, the façade design phase and the natural light analysis phase. The façade is validated in three types of building envelope, that is, kinetic façade, static façade, and building without a façade (only glazed façade) through testing with the *DIVA* software that analyses in terms of natural light. This research focuses on glare analysis to improve visual quality in the space. *Rhino* and *Grasshopper* computational methods were used to design façade optimisations for the simulation. *Rhino* software is used to build 3D forms while *Grasshopper* software uses two plug-ins - *DIVA* to analyse natural light and *Wallacei* to generate façade patterns. The kinetic façade integrates with biomimicry science to design the physical and behavioural components. The physical component uses characteristics of DNA structure to generate probability patterns and phototropism behaviours of plants, and is applied in the façade movement to respond to sunlight. Then, both inspirations are interpreted into façade patterns to generate probability patterns in *Wallacei* software and the potential of each façade type is analysed in *DIVA* software to optimise for the glare factor.

3.1. Comparing effectiveness

Data was collected to compare the effectiveness of building envelopes with a static façade, kinetic façade, and without a façade to simulate the use of space in the same zone by focusing on the glare factor. From Figure 5, the variables are as follows:

3.1.1. Independent variables

- The building envelope:
  (1) Without façade (only glazed panels).
  (2) Static façade.
  (3) Kinetic façade.

3.1.2. Dependent variables

- Natural light, focused on the glare factor.

3.1.3. Control variables

- No existing buildings nearby, same zone for simulation.
- Location (Bangkok, Thailand), climate (clear sky), period (Winter Solstice – 22 December), time (7.00 AM – 5.00 PM).

4. Results and discussion

4.1. Analyse potential façade patterns (Phase 1)

This section analyses potential façade patterns inspired by physical DNA and phototropism behaviour to construct the strip form. *Wallacei* is used to generate potential façade patterns, which is an evolutionary algorithm software to generate algorithms forms. Evolving algorithms are a series of optimisation algorithms that imitate the mechanisms of natural evolution [7].

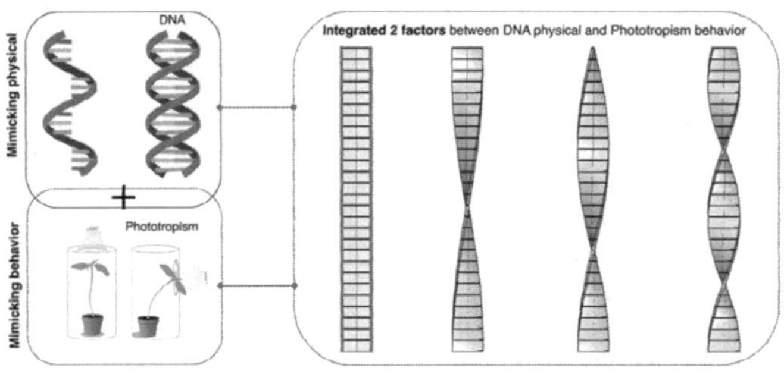

Figure 6. A façade inspiration that integrates physical DNA and phototropism behaviour.

Figure 7. The environment affects the faces of birds of the same species.
Photograph: Ann Ronan Pictures/Getty Image

Figure 6 shows the processes that mimic strip DNA. This study mimics two parts for a flexible movement purpose; the physical DNA structure and phototropism behaviour. Phototropism is most commonly found in plants and involves the development of an organism in response to a light stimulus (external factor) by photoreceptors related to signalling mechanisms. According to external factors, plants adaptively change their growth characteristics according to the amount of accessible sunlight [8]. This mechanism is crucial for interpreting the façade strip mechanism since it must adjust itself to the sunlight, similar to plants.

Phenotypes are measurable traits of an individual that derive from the association of its genotype (total genetic inheritance) with the environment. Features that may be observed include behaviour, biochemical properties, colour, shape, and size, for example the bird species depicted in Figure 7 [9]. Thus, the phenotype's simple definition are external appearances or characteristics that are easy to observe.

Figure 8(a) shows the strip phenotype probability patterns, resulting in 450 solutions from generation 0 to generation 8. Each generation has different and repetitive patterns compared to other generations. For instance, generation 0 and generation 5 have some strips with the same patterns. The strip patterns that occur from the phenotype are different, resulting in various strip phenotypes. The strip phenotype occurs from mimicking DNA strip and phototropism phenomenon to simulate the characteristic of strip patterns for application in the kinetic façade model. The generation of new solutions was stopped after 450 solutions had been generated, since when looking at the standard deviation trend graphs (Figure 8[b]), the final generation had a smaller slope than the first generation. This means that the strip phenotype repeated the same patterns and has no benefit for the purpose of this research. Thus, the final generation at which the generation process was paused was generation 8. A low standard deviation factor indicates that most values are similar to each other (less variation within the population). In contrast, a high standard deviation means that the values are farther from the mean (greater variation within the population) [10]. When looking at the standard deviation graph, early generations had more varied strip patterns than the final generation, as Figure 9 shows the probability output in 50 solutions of strip phenotype in generation 3, indicating greater repetition of strip patterns than the early generation. However, it can be seen that this generation has various strip phenotypes when observing the twist angle characteristics that respond to sunlight.

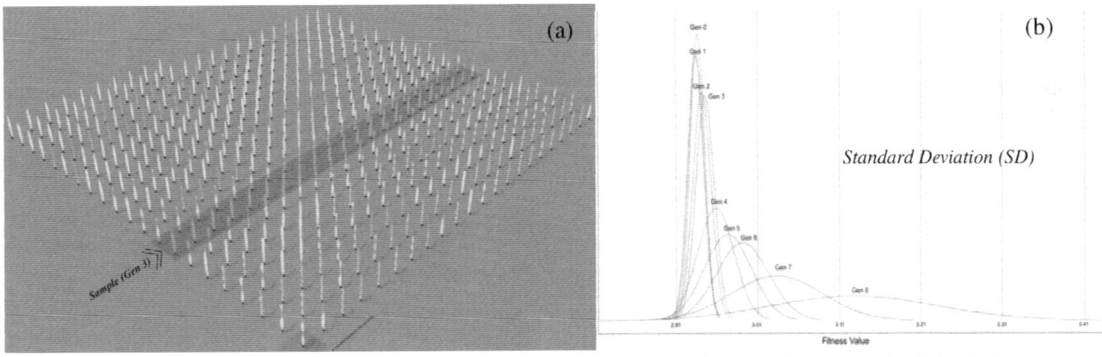

Figure 8. Overall generation of the strip phenotype in 450 solutions and the standard deviation trend.

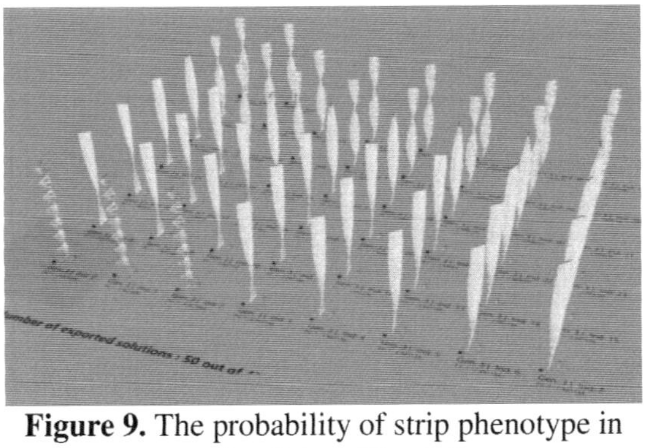

Figure 9. The probability of strip phenotype in generation 3 into 50 solutions.

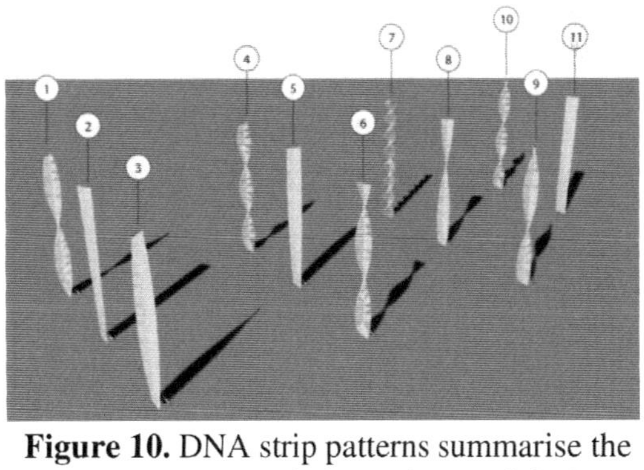

Figure 10. DNA strip patterns summarise the different characteristics and potential after simulating with sunlight.

Figure 10 summarises the strip patterns from generation 0 to generation 8 that were applied in the kinetic façade model. These patterns were selected since they respond to sunlight differently, similar to phototropism behaviours. The various angles of the strip patterns have differing potential to block sunlight at different times. Thus, it is beneficial for adapting to user needs in various situations when living in spaces. When there is less light intensity, the strip will be more twisted than the high light intensity, similar to a strip 7 as in Figure 10.

Figure 11 presents the probability of façade patterns in generation 0. The solutions that the researcher exported is 50 out of 1150 solutions. Indeed, the phenotype generates into 23 generations, including generation 0 to generation 22. When comparing the kinetic façade from generation 0 to generation 22, the results are similar to generating DNA strips when considering standard deviation, fitness values, and standard deviation value trendline which are consistent. Therefore, this output can predict that the kinetic façade can twist or adjust to sunlight, similar to the strips in Figure 10.

Figure 12 shows the graph trend that can predict the possibility of the kinetic façade characteristic. First, standard deviation represents the distribution of a set of values from the mean [11]. The chart's objective is to present and analyse the levels of variation and convergence for each generation in the population. Increased variation is represented through a 'flat' curve, while increased convergence is represented through a 'narrow' curve. Thus, it can predict that each generation's façade patterns are independent since each generation's curve is similar. Nevertheless, each generated façade has a range of patterns, for instance the early generation's façade patterns have various phenotypes when compared to the last generation. Second, the fitness values chart analyses the fitness values for each fitness objective independently across the entire population. The aim is to visualise how the solutions perform in relation to one another, both within each generation and across the population. When analysing the graph, most of the generation is independent since most line graphs are relatively stable. However, the early generation has a fluctuating line which means that the phenotypes or external appearance are more diverse than the stable line. Third, the standard deviation trendline chart presents the standard deviation value for each fitness objective independently, and for each generation across the entire simulation from start to finish. The aim is to highlight specific trends in the variation and convergence of each generation across the population. Thus, after analysis, the graph has a similar meaning to the standard deviation graph. This illustrates that the early generated patterns were more varied than the last generation or the early generation has greater diversity than the last generation.

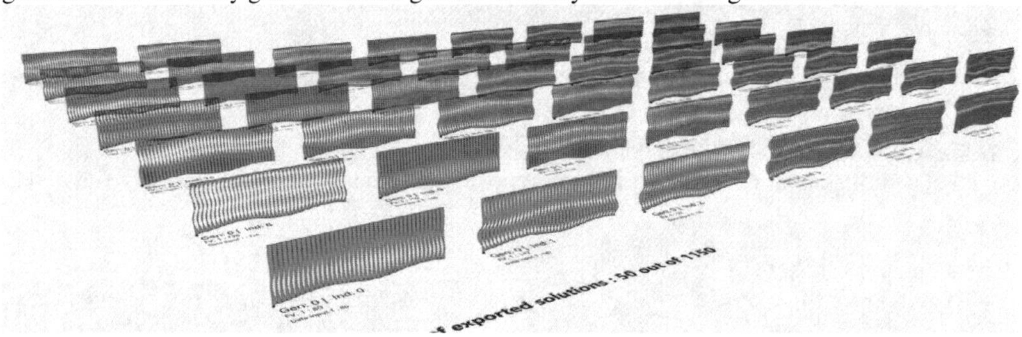

Figure 11. Phenotype of façade generation in generation 0 (50 out of 1150 solutions).

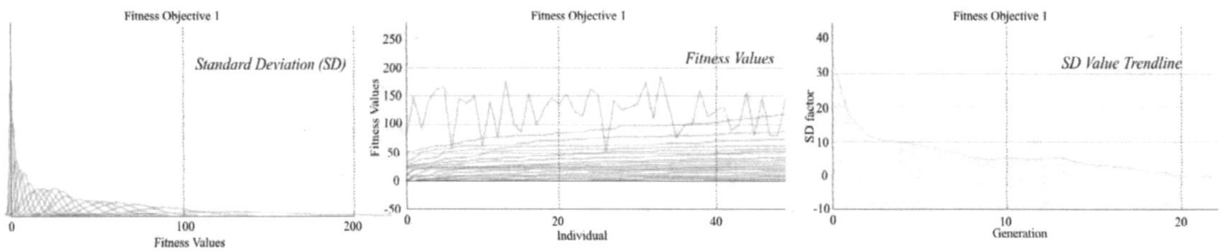

Figure 12. Trend of façade characteristics.

4.2. Analysis of natural light ambient (Phase 2)

This part analyses the building envelope in working spaces with a static façade, kinetic façade, and only glazed panels (without façade). The kinetic façade refers to cases generated in phase 1 to simulate effective glare protection. Figure 13(a) illustrates the setting model, which is the working space for simulation by the node. The node or dot is crucial for calculating the glare factor, hence the researcher set the node position close to the human eye at a distance of approximately 1.10 metres when doing work activities as in Figure 13(b).

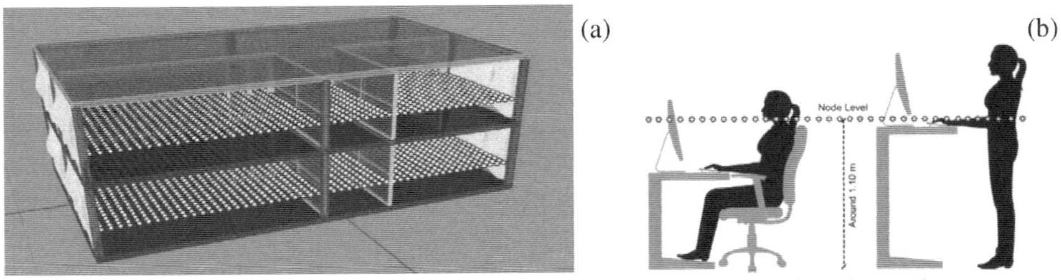

Figure 13. Model setting to simulate the glare factor using the node.

4.3. Glare analysis

Table 1. Identification of occupant sensations of discomfort.

DGP classification	Glare range values
Intolerable Glare	DGP ≥ .45
Disturbing Glare	.45 > DGP ≥ .4
Perceptible Glare	.4 > DGP ≥ .35
Imperceptible Glare	.35 > DGP

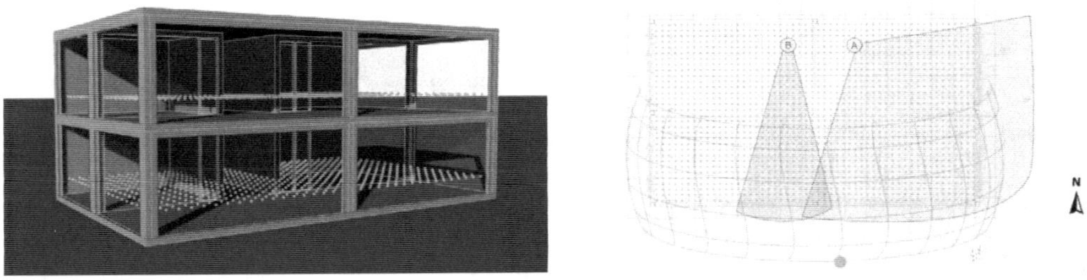

Figure 14. Building model without a façade and the zone to simulate both views A and B.

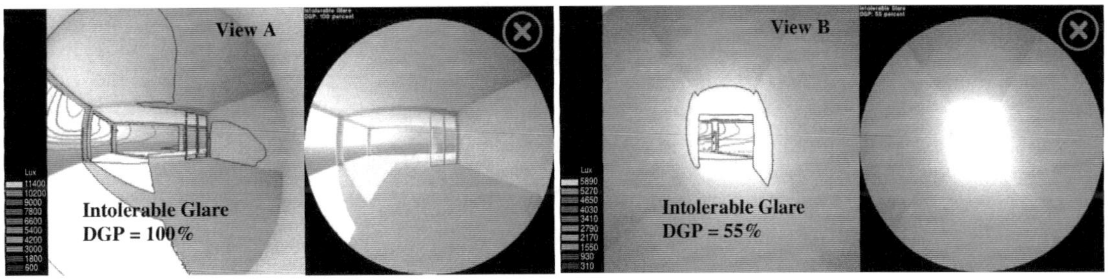

Figure 15. Glare analysis with a building model (without a façade).

Glare occurs when part of an interior space is significantly brighter than the room's luminosity. Harm can occur when the eyes are exposed to a more intense light than they are normally adapted to. The most prevalent cause of glare arises from the light source and window being in the same visual area, either directly, through reflections, or both simultaneously [12]. Daylight glare probability (DGP) was established in a new glare prediction model since existing glare indices did not accurately estimate the degree of daylight glare irritation in a working atmosphere with regular work practices and diverse non-uniform sources of glare, such as the Venetian blind system [13], [14]. GDP was derived from laboratory results in private office spaces containing human test subjects. Table 1 shows a sensation of discomfort for users of such spaces, including DGP classification and glare range values. The glare classifications are as follows: intolerable glare; disturbing glare; perceptible glare; and imperceptible glare [15]. Glare types that negatively affect humans are intolerable glare and disturbing glare. Thus, designers should avoid and protect against these types of DGP since they will occur if spaces have an exaggerated open space or glazed façade.

Figure 14 shows the 3D model and zoning for simulating the glare factor for views A and B. These zones were selected since they have different light intensity, with view A receiving greater light intensity than view B which is a narrow space. The simulation must control the input data as follows: location; climate; and period. This study is based in Bangkok, Thailand, with clear skies and during the Winter Solstice period (22 December) between 7.00 AM and 5.00 PM. During the Winter Solstice period in Thailand, the building can receive more direct sun than during other periods, making it suitable for testing.

Figure *15* illustrates a glare analysis with a building model without a façade. For View A, the result is intolerable glare and the daylight glare index versus the percentage of persons disturbed (DGP) is 100%. View B shows the glare analysis in another area, which resulted in intolerable glare with a DGP of 55%.

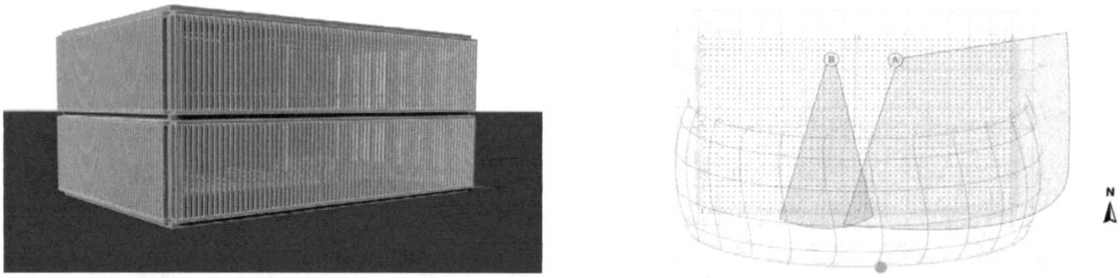

Figure 16. Model with a static façade and zone to simulate views A and B.

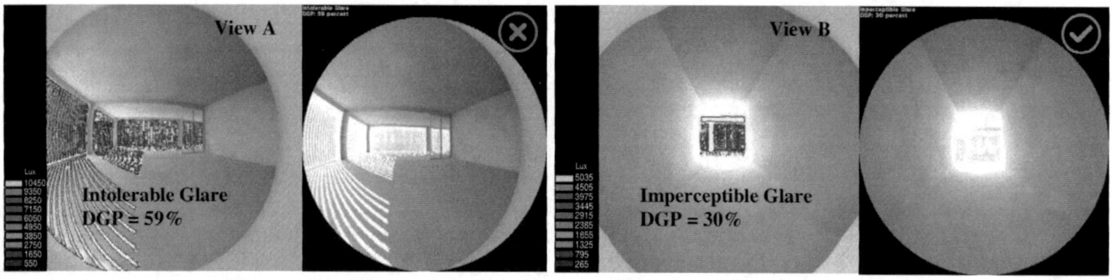

Figure 17. Glare analysis with a building model (static façade).

Figure 16 shows the model and the zone of both views A and B to simulate the glare factor. In Figure 17, view A shows the glare analysis with the static façade building model. The result is intolerable glare and a DGP of 59%. View B shows glare analysis, with the result being imperceptible glare and a DGP of 30%. Thus, the potential of a static façade is useful for some areas that are far from glazed panels.

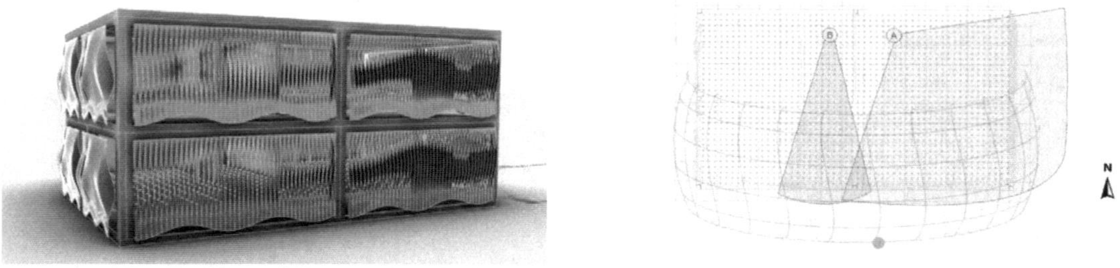

Figure 18. Building model of a kinetic façade and zone to simulate views A and B.

IOP Conf. Series: Materials Science and Engineering 1148 (2021) 012002 doi:10.1088/1757-899X/1148/1/012002

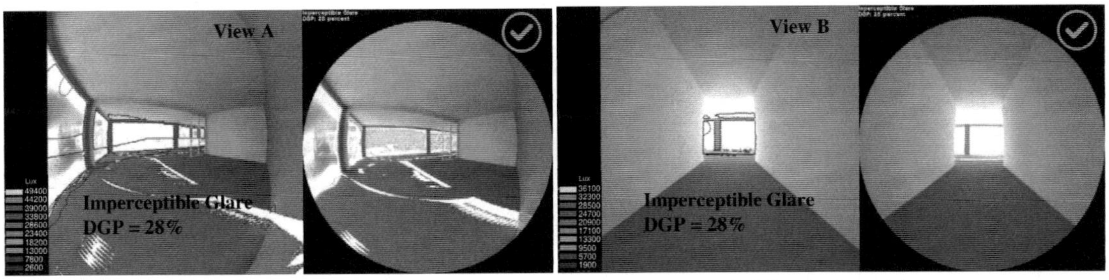

Figure 19. Glare analysis with a building model (kinetic façade).

Figure 18 shows the model and the zone of views A and B to simulate the glare factor. In Figure 19, view A shows the glare analysis with the kinetic façade building model, which indicates imperceptible glare and a DGP of 28%. View B illustrates glare analysis with the result of imperceptible glare and a DGP of 28%. Figure 20 illustrates a comparison of the three-building envelopes in terms of glare factor. The result is that a kinetic façade can better protect against the glare factor than a static façade, which is useful in some areas and inappropriate glare in some periods (not stable glare factor – view B).

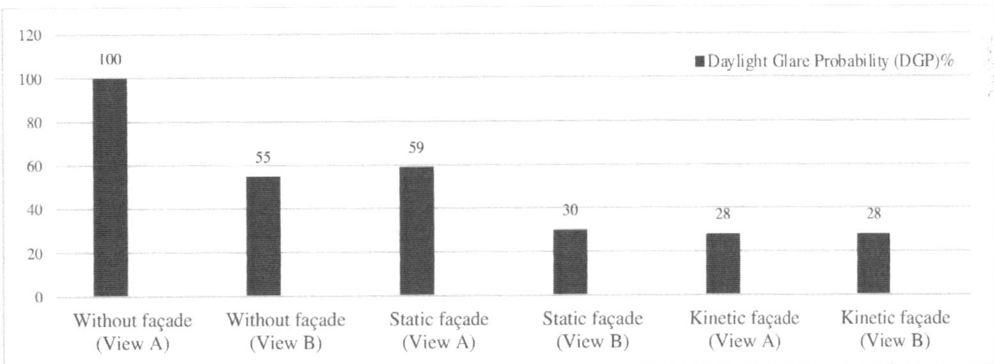

Figure 20. Comparing the potential of the three-building envelopes in terms of glare factor.

5. Conclusion

One crucial factor to encourage users to work effectively in working spaces is the suitable glare factor, since this can negatively affect people and reduce their working effectiveness. Thus, this study focused on simulating the potential of a kinetic façade by comparing it with a static façade and a building without a façade (only glazed panels). The kinetic façade uses biomimicry science to attain a façade pattern by mimicking nature, both in physical appearance and behaviour. The kinetic façade was inspired by nature, including DNA structure and the phototropism behaviour of plants. Thus, these inspirations must interpret a physical appearance for simulation in the evolution stage to discover the potential façade patterns by using *Wallacei*. After simulating the façade patterns, the next process simulated each building envelope's effectiveness in the preparing space, from views A and B (
Figure *15*, Figure 17, and Figure 19). First, the DGP of the building without a façade from views A and B was 100% and 55%, indicating intolerable glare and unsuitability for use as a space. Second, the DGP of a building with a static façade from views A and B was 59% and 30%, with view A having intolerable glare and view B subject to imperceptible glare, indicating that the space is suitable for users only from view B (narrow space). Third, the DGP of a building with a kinetic façade from views A and B was 28%, with both views subject to imperceptible glare. Hence, both areas are suitable as working spaces. The building with a kinetic façade has the potential to protect against glare. Furthermore, if the façade receives less natural light, it will twist itself to receive outdoor light following the same pattern as the seventh strip in Figure 10. Thus, the kinetic façade is a crucial element that should be installed in buildings with exaggerated proportions of open spaces and glazed panels.

6. References

[1] Chaloeytoy K, Ichinose M, Chien S-C. Determination of the Simplified Daylight Glare. Buildings. 2020;10(180).

[2] Kubba Sam. Handbook of green building design and construction: LEED, BREEAM, and Green Globes. Butterworth-Heinemann; 2017.

[3] Facchin A, Daini R, Zavagno D. The Glare Effect Test and the Impact of Age on Luminosity Thresholds. Frontiers in Psychology. 2017;8:1132.

[4] Christoph Reinhart, Shelby Doyle JAJ and RM. Glare Analysis - Recommendations for Practice [Internet]. [cited 2021 Jan 12]. Available from: http://web.mit.edu/tito_/www/Projects/Glare/GlareRecommendationsForPractice.html

[5] Katabaro JM, Yan Y. Effects of Lighting Quality on Working Efficiency of Workers in Office Building in Tanzania. Jia C, editor. Journal of Environmental and Public Health. 2019;2019:3476490.

[6] Government of Canada CC for OH and S. The Young Workers Zone : Teaching Tools : Physical Hazards: Lighting. 2016.

[7] Alateeq M, Pedrycz W. A comparative analysis of bio-inspired optimisation algorithms for automated test pattern generation in sequential circuits. Applied Soft Computing. 2021;101:106967.

[8] Goyal A, Karayekov E, Galvão VC, Ren H, Casal JJ, Fankhauser C. Shade Promotes Phototropism through Phytochrome B-Controlled Auxin Production. Current Biology. 2016;26(24):3280–7.

[9] The Editors of Encyclopaedia Britannica. Phenotype | genetics | Britannica [Internet]. September 05. Encyclopædia Britannica; 2016 [cited 2021 Jan 11]. Available from: https://www.britannica.com/science/phenotype

[10] Jim Frost. Measures of Variability: Range, Interquartile Range, Variance, and Standard Deviation - Statistics By Jim [Internet]. [cited 2021 Jan 17]. Available from: https://statisticsbyjim.com/basics/variability-range-interquartile-variance-standard-deviation/

[11] The Standard Deviation [Internet]. [cited 2021 Jan 17]. Available from: https://www.le.ac.uk/oerresources/ssds/numeracyskills/page_17.htm

[12] Fagerhult. Glare - Fagerhult (International) [Internet]. [cited 2021 Jan 17]. Available from: https://www.fagerhult.com/knowledge-hub/EN-12464-1/glare/

[13] Wienold J, Christoffersen J. Evaluation methods and development of a new glare prediction model for daylight environments with the use of CCD cameras. Energy and Buildings. 2006;38(7):743–57.

[14] Konstantzos I, Tzempelikos A, Chan Y-C. Experimental and simulation analysis of daylight glare probability in offices with dynamic window shades. Building and Environment. 2015;87:244–54.

[15] Sun Y, Liu X, Qu W, Cao G, Zou N. Analysis of daylight glare and optimal lighting design for comfortable office lighting. Optik. 2020;206:164291.

Acknowledgements

This research was supported by the research fund of Japan Advanced Institute of Science and Technology (JAIST), Japan. The research fund of Sirindhorn International Institute of Technology, Thammasat University, and National Electronics and Computer Technology Centre (NECTEC), Thailand.

IOP Conf. Series: Materials Science and Engineering 1148 (2021) 012003 doi:10.1088/1757-899X/1148/1/012003

The utilization of green roofs and walls "ecosystem services" as a strategy to mitigate climate change

Susana Moya Vicuña[1*] **and Marcelo Villacis Ormaza**[1, a]

[1]Faculty of Architecture, Arts and Design (FAAD), Universidad Tecnológica Indoamérica, Sabanilla Street, Quito 170103, Ecuador
[a]marcelovillacis@uti.edu.ec

Abstract. Green infrastructure is a set of systems and strategies that promote a more sustainable and efficient development, this paper reviews the green roofs and facades that are part of this infrastructure. Currently, its application in buildings aims to take advantage of its "ecosystem services", such as energy savings, reduction of the heat island, improvements in water quality, air and health, and other advantages that are broken down into the article, which are achieved through natural processes carried out by vegetation, which allows evapotranspiration, the absorption of impurities, the reduction of noise, the capture of rainwater. Thus, collaborating in the reduction of sewage systems load in the cities, thanks to the use of the water captured, also contributing to water cost reduction from the drinking network of the cities. This paper results from a literature review on advantages and disadvantages that green roofs and walls offer in an urban environment.

1. Introduction

Architecture is dominated by an aesthetic and globalized idea, in which the environmental impact or energy expenditure was not considered, triggered a series of phenomena such as the heat island that contribute accelerating climate change and affect all living beings in our planet.

However, since these phenomena have begun to affect the quality of life of human beings in an increasingly notorious way, an ecological thought has been awakened influencing disciplines such as architecture. In 2019 the European Union describes green infrastructure as a *"strategically planned network of natural and semi-natural areas with other environmental features designed and managed to deliver a wide range of ecosystem services*" [1], among these we find the green roofs and walls that integrate the natural environment in the architecture and which we will deal with in this article.

The integration of vegetation in architecture is not something new, this synergy has occurred throughout history in different cultures around the world with the purpose of protection, adaptation to different climates and as crops solution.

Green roofs, as we know them, were originated in the early 20th century in Germany, where vegetation was installed on roofs to mitigate the damaging effects of solar radiation on the roof structure, coinciding with a growing concern for the environment appeared in the 70s, especially in urban areas, which together with a large number of studies and regulations developed in German cities, allowed the green roof technology to be quickly adopted in terms of facades [2], vegetation has been present through climbing plants in various buildings, especially in Germany and England, where interventions were carried out on buildings using vegetation on the façade [3].

This article will analyse the advantages and disadvantages that green roofs and walls offer in sociological, environmental, functional, energy and comfort aspects, by collecting and analysing numerous state-of-the-art references on the topic.

2. Green infrastructure

In 2019 the European Commission describes Green infrastructure as *"A strategically planned network of natural and semi-natural high-quality areas, designed and managed to provide a wide range of ecosystem services and protect the biodiversity of rural and urban settlements. has for objective improve the nature capacity to provide multiple and valuable goods and ecosystem services, such as clean water or air."* [1] in some cities can find this infrastructure as Green roofs and walls, used to generated ecological corridors, that provide ecosystem services, but its efficiency depends of the characteristics of its substrate, vegetation and technology.

2.1. Green walls

There is no established definition, however, we can say that it´s one that allows the integration between the vegetation and architecture, by partially or totally covering vertical surfaces indoors or outdoors and that we could generally differentiate into two groups, Table 1 [4]-[6].

Table 1. General characteristics of Green walls.

	Kind	Investment	Maintenance	Density/ height vegetation	System
Green Wall	Traditional	low	low	medium	Extensive
	Vegetable curtain or double skin	medium	low	medium	Extensive
Living walls	Panel	high	high	medium/ high	Intensive
	Geotextile felt	high	high	medium/ high	Intensive
	Perimeter planters	medium	high	medium/ high	Intensive

2.2. Green roofs

It is Vegetal coverage, installed in a horizontal area such as a building roof, allowing the integration between vegetation and architecture and usually has a similar basic configuration, however, we can differentiate into two groups, Table 2 [7], [8].

Table 2. General characteristics of Green Roofs.

Type	Investment	Maintenance	Substrate	Use	Density an height of vegetation
Extensive Green Roof	low	low	slim	generally ecological	low/ medium
Intensive Green Roof	high	medium/high	heavy	generally aesthetic	Medium/ high

3. Ecosystem services

"Ecosystem Services" concept was born, for the first time, in 1977, under the name of "services of nature", when Westman suggested cataloging the functions of the ecosystem as services in order to increase interest in biodiversity conservation, this concept generally include all the resources produced by nature, which are used by human beings directly or indirectly, and improve or favor our survival and life quality [9].

3.1. Air quality improvements

Human beings, as well as many of our activities, including construction processes, produce a carbon footprint (CO2), in many cases even causing deforestation.

Vegetated roofs and facades contribute to climate change mitigation, directly removing CO2 from the atmosphere through photosynthesis, based on several annual studies suggesting that urban areas would be neutral to their carbon production, if its natural fraction was greater than 80%, which was corroborated by M. Demuzere in 2014 [10], [11].
Vegetation also captures particles and environmental dust, which adhere to leaves, to be later eliminated with the rain, thus cleaning the air, also according to an analysis carried out by Demuzere in 2014, entitled *"Mitigating and adapting to climate change: Multi-functional and multi-scale assessment of green urban infrastructure"* which, although there is little evidence available, suggests the possibility that vegetation, depending on its type and size, is capable of absorbing pollutants such as particles (PM10), black carbon and other short-lived weather pollutants [11], other research in cold climates like the one present in eastern Europe showed that, pollutants such as NO2 and O3 are clearly reduced [12], while in other areas like in the northeast of China, NO2 levels do not have a significant variation with the increase or reduction of vegetation, but O3 will behave differently, resulting in lower levels with a higher density of vegetation [13].
It is important to consider that, the capacities of green roofs and facades to eliminate environmental pollutants are linked to their location, meteorological factors, and to the correct choice of vegetation, for example, several research have shown that the concentration of pollutants is inversely proportional to the quantity of urban vegetation, which helps improving air quality, but the pollutants on which it acts will vary even by minor factors like the type of leaf vegetation, so it is difficult to stablished stable parameters whose application ensures a decrease in environmental pollutants, so each sector must establish its own parameters that allow it to take advantage of this ecosystem service [14], [15].

3.2. Temperature regulation - reduction of urban heat island effect and energy saving

The warming of cities can have consequences that range from the increase in the use of electricity due to the consumption of air conditioning, to the appearance of health problems for inhabitants, Green roofs and walls collaborate in temperature regulation. Green facades are efficient in generating shade for buildings, vegetation absorbs with its leaves part of the radiation to be used in the photosynthesis process and reflects the other part to the atmosphere, which generates a better indoor climate, with the consequent reduction of the use of energy, although as mentioned Briony and Norton in 2015, in a study carried out in Melbourne - Australia, they are not as efficient when it comes to provide shade at street level, however it significantly benefits the cooling of the city, since it captures and returns the solar rays preventing these from heating the surfaces [16].
Considering that, during the day, roofs are some of the hottest surfaces in urban areas, the use of green roofs offer great possibilities in the regulation of temperature, and so energy savings, as shown in a research carried out in 2016 by Gómez. M. et. al, where, through simulations and analysis of existing and modeled roofs in Portugal, concluded in general terms the influence on the annual energy expenditure that parameters such as the thickness of the substrate have, with an incidence capacity of 115%, the density and height of vegetation with an incidence of 38 and 7% respectively and irrigation, tested in intensive systems, with flow changes from 0 mm / day to 6 mm / day showing an incidence of around 500%[17], considering the variation in energy consumption that different combinations of these parameters pose, the choice of a suitable green roof for each environment becomes relevant.
According to several research, intensive roofs with taller and denser vegetation would be more efficient in hot or warm climates, where the highest waste of energy is for cooling, considering that the greater the vegetation the greater the evapotranspiration, which consists on the absorption of water through the roots of the plants and the emission of this through their leaves, by taking advantage of the heat present in the atmosphere to evaporate water providing moisture, which generates cooling of the surrounding air, thus reducing the temperature of the ground surface and the flow of heat through the roof and facades by absorption of radiation, collaborating not only with the reduction of energy use for cooling, but also with the reduction of the heat island effect, on the other hand, irrigation requirement will be greater cooling the covers [17]-[21].
In 2008, Alexandri and Jones study of *"Temperature decreases in an urban canyon due to green walls and green roofs"* in 9 cities around the world showed that the hotter and drier the climate, more important the effect of green walls and roofs in mitigating temperatures [22]. In addition, energy

savings of 32% to 100% can be achieved in the cooling of buildings, regarding green facades [23], [24], concluded that the application of grass as vertical green reduced the temperatures of the interior surface by more than 2 ° C [11].

3.3. Reduction of flood problems and water quality improvement

Cities are covered with impermeable surfaces that are part of their buildings, roads, and even parks and squares, which causes alterations in the ways in which rainwater flows. In a natural environment in which there are vegetation and permeable surfaces, rainwater would be captured and used by the vegetation, as well as filtered by earth to become part of groundwater, however, in cities this does not happen and is it translates into an increase in surface water runoff that can cause flooding due to saturation of the sewage system, in addition to degeneration and waste, due to polluting particles drag [25]. As has been well documented in Europe and the United States, it is green roofs that, due to their location in buildings, as well as their configuration, collaborate with water retention, since they have a layer in which rainwater is collected and stored, "being able to retain considerable amounts of rain by reducing the annual volumes of runoff in areas where rainfall is high, unlike climatic zones characterized by summer droughts and extended winter wet seasons" [18], where this capacity is not a relevant issue seen from the point of runoff, but it´s important from the point of water storage for self-sufficiency, especially in drought times.

According to several studies, green cover improves water quality, eliminating various pollutants such as phosphorus by 60% and fecal coliforms by up to 100%, according to tests by Gregoire and Clausen [26], reducing pollutants global load, since that act as a sink. Its removal efficiency will depend on the type of pollutant, as well as vegetation type, soil properties, fertilizer addition, and local climate [11], factors that can become very important when measuring pollutants in the water stored by green roofs.

Other research like the one carried out with extensive roofs in Berlin in 2014, in which it was intended to reduce pollution in the Spree river, the one carried out in Poland in 2018 or the one carried out in Colombia in 2020 in varied climates, and in which they experimented with substrates of different compositions showing that green roof substrates can be an important source of pollutants such as phosphorus, TSS, TKN, PO4, DQO, DBO, nitrogen and coliforms in runoff water, due to the ability of the substrates to remove these pollutants are affected by fertilization, organic matter, pollutants in the substrate, on the roof membranes, the thickness of the substrate, or the type of vegetation.

Additionally an intensive cover with deep substrates and higher vegetation would improve its capacity to capture water, however, if in addition to being intensive the substrate is old, it may have a higher concentration of pollutants than “newer” subtracts or specifically treated to achieve that the chemical properties of its organic and inorganic components, as well as its plant species, work to achieve a cleaner runoff, therefore, the selection of plants and substrates that optimize the absorption of nutrients and pollutants can help improve the capacities of the roofs in relation to water quality, always taking into account the climatic and natural characteristics of the environment, although they can affect other ecosystem services such as the ability to save energy in ventilation heating [20], [24], [27], [28].

3.4. Acoustic isolation and urban noise reduction

The ability of vegetation to isolate us from noise, by fences that surround the buildings, has been well known practically since the construction of buildings began. In today's cities that are increasingly filled with hard surfaces that replace vegetation, we find ourselves exposed to high levels of noise, an increase in green infrastructure in cities can reduce these noise levels since while hard surfaces of urban areas tend to reflect sound, vegetation and the substrate that make up the roofs and vegetated facades can absorb it. Vegetation works by reducing sound levels in three ways:

- Vegetable elements such as trunks, branches, and leaves can reflect and disperse noise.
- By the acoustic absorption of vegetation due to its mechanical vibrations caused by sound waves, which lead to dissipation by converting sound energy into heat.
- The reduction of sound levels due to the interference of acoustic waves in the substrate layers [29].

Research carried out to measure noise reduction levels in green infrastructures determined that “noise reduction was lower for green facades than for roofs” [30].

A comparison between a roof made of typical materials and a green roof made by Rowe determined that because the substrates used for green roofs are (highly) porous, they allow a large number of interactions between sound waves and substrate particles [31], which produces a noise attenuation, with a variation that will depend on substrate thickness, since increasing its depth up to 15 to 20 cm, improves the reduction of noise in the medium frequencies and the vegetation in the high frequencies, so that an increase in the thickness of vegetation or the use of species with more foliage could increase its acoustic insulation capabilities, however, the type of support structure and the materials to be used and the fact that there must be a physical separation between construction elements to avoid the transmission of sound [29].

3.5. Ecosystems recovery

Vegetated roofs and facades have the ability to "imitate or recreate" natural ecosystems, particularly the extensive roofs that mimic typical habitats, attracting biodiversity to these spaces.

An investigation carried out by Braaker and others, on the ecological and functional value of green roofs in cities, according to the variation of biodiversity of four groups of arthropods with contrasting mobility in 40 green roofs and 40 extensive green sites on the ground in the city of Zurich, Switzerland, the study revealed that the connectivity of the habitat between surrounding green roofs promotes the exchange of insects and pollinators between the roofs, especially in regard to those with high mobility, while the less mobile groups are more influenced by local environmental conditions" [32], however, greater connectivity between green spaces could eventually even promote the exchange of low-mobility insects, collaborating with the ecological regeneration and repopulation of insects such as bees currently in danger. This can give us a reference for how effective they are for the development of biodiversity since there are very few studies on this subject.

Nevertheless, wildlife is not always desirable, especially if the green roof is not designed to support biodiversity especially of animals such as birds or rodents, which could cause damage to the elements that make up the roof, or its maintenance. In addition, it should be considered that these spaces are relatively isolated from larger natural green spaces, due to the difference in height so that, although they collaborate in mitigating the lack of natural green spaces, they cannot be considered as a replacement, at least not from a biodiversity point of view of medium or large animals [18].

3.6. Green recovery and quality of life improvements

These green infrastructures contribute to the increase of urban flora and fauna, generating relaxing visual spaces, with great potential for the development of recreation and relaxation activities such as urban agriculture, contemplation of nature, meditation, etc. Exposure to these "natural" environments provide psychological benefits, reducing stress, crime, insecurity feeling, and even improving physical health, or contributing to economic and food independence as several studies have shown and which is reflected in the requirements to be met in several sustainable constructions.

For Miller and Charlie, forty years' experience in Germany suggests that they may have value in the protection of waterproofing materials, since the multiple layers of the green roof could protect them from mechanical damage, ultraviolet radiation, and extreme temperatures, minimizing damages caused by the daily expansion and contraction of roofing materials [12], [18].

4. Discussion

Intensive and extensive Green roofs and walls systems provide a wide range of environmental benefits through ecosystem services, Table 3 shows a resume of the environmental benefits of intensive and extensive systems, however it is important to understand that to determine the most effective system to be used, first must be know the climatic and even constructive conditions of the environment, therefore, it cannot be taken as a base applicable to all environment, and is susceptible to changes according to the particular characteristics of the vegetation, the type of substrate, fertilizers, pollution levels, climate, among others.

Table 3. Resume environmental benefits of intensive and extensive systems.

System	Air Quality Improvements / Temperature Regulation	Water quality improvement / flood	Acoustic Isolation and Urban Noise Reduction	Green Recovery and Quality of Life Improvements
Extensive	high air permeability low evapotranspiration low shade possibilities Medium radiation absorption high pollutant removal	low capacity in water collection low water consumption	low reflection and dispersion low Isolation and absorption low destruction of acoustic waves	Increase in biodiversity Improvements in quality of life and health
Intensive	low air permeability high evapotranspiration high shade possibilities high radiation absorption low pollutant removal	High capacity in water collection high water consumption	low reflection and dispersion low Isolation and absorption low destruction of acoustic waves	Increase in biodiversity Improvements in quality of life and health

5. Conclusions

The use of vegetated roofs and facades can collaborate with climate change effects mitigation, and their contributions will be greater the more widespread they are since their ecosystems not only act on the building in which they are installed but also their benefits extend to the environment, so its use in cities could be used to solve various problems such as pollution caused by industry, the high number of vehicles; the risks of flooding due to climatic factors, weaknesses in its sewage systems; or even combat the heat island.

However, one of the factors in which their collaboration, although little documented is of great importance, is in the recovery of ecosystems and green spaces that not only improve the quality of physical and mental life of human beings, but they collaborate with the recovery of habitats for urban fauna and even promote pollination, which currently occupies environmentalists and that can be solved through the implementation of policies that promote the use of green infrastructure, which has begun to be seen in some cities such as Toronto, Sydney, Copenhagen, and others around the world where there are action plans to preserve it.

6. References

[1] Comisión Europea 2019 *Guidance on a strategic framework for further supporting the deployment of EU-level green and blue infrastructure* (Brussels, Belgium SWD) p 1

[2] Oberndorfer E, Lundholm J, Bass B, Coffman R, Doshi H, Dunnett N, Gaffin S, Köhler M, Liu K Y and Rowe B 2017 Green roofs as urban ecosystems: ecological structures, functions, and services. *BioScience.* **57** 823-833.

[3] Gerot M 2014 *Muros y Fachadas Verdes, Sistemaas y Plantas, Funciones y Aplicaciones*, ed Icaria (Madrid Spain) pp 15-30.

[4] Hadba L, Silva L, Mendonça P 2017 *Conf. Living and Sustainability: An Global Critique of Desing and building Practices,, Localu and global.* (London England) (London South Bank University) vol. 1 pp 241-251

[5] Pérez , Rincon L, Vila A, Gonzalez J and Cabeza L 2011 Green vertical systems for buildings as passive systems for energy savings. *Applied Energy*, **88** 4854-4859

[6] Loh S and Stav Y 2008 *Proceedings of the Subtropical Cities* prints.qut.edu.au (Breisbane, queesland Australia)

[7] Minke G 2005 *Techos Verdes, Planificación, Ejecución Consejos Practicos* ed fin de siglo (Montevideo Uruguay) pp 32-51
[8] Yeomans F, Almada D and Martínez R 2013 *Innovation in Engineering, Technology and Education for Competitiveness and Prosperity*, LACCEI (Cancun Mexico)
[9] Gómez E, De Groot R, Lomas P L and Montes C 2009 The history of ecosystem services in economic theory and practice: from early notions to makets and payment schemes. *Ecological Economics.* **69** 1209-1218
[10] Nordbo A, Järvi L, Haapanala S, Wood C R and Vesala T 2012 Fraction of natural area as main predictor of net CO2 emissions from cities. *Geophysical Research Letters.* **39** L20802
[11] Demuzere M, Orru K, Heidrich O, Olazabal E, Geneletti D, Orru H, Bhave A G, Mittal N, Feliu E, Faehnle M 2014 Mitigating and adapting to climate change: Multi-functional and multi-scale assessment of green urban infrastructure. *Environmental Management.* **146** 107-115
[12] Yli-Pelkonen V, Setälä H and Viippola V 2017 Urban forests near roads do not reduce gaseous air pollutant concentrations but have an impact on particles levels. *Landscape and Urban Planning.* **158** 39-47.
[13] Yli-Pelkoen V, et al, 2018 The impact of urban trees on concentrations of PAHs and other gaseous air pollutants in Yanji, northeast China Atmospheric *Environment* **192** 151-159
[14] Badach J, Dymnicka M, Baranowski A 2020 Urban vegetation in air quality management: a review and policy framework *Sustainability.* **12(3)** L1258
[15] Suszanowicz D and Wiecek A 2019 The impact of green roofs on the parameters of the environmente in urban areas – review *Atmosphere* **10** L792
[16] Norton B, Coutts A, Livesleya S, Harris R, Hunter A M, Williams N 2015 Planning for cooler cities: a framework to prioritise green infrastructure to mitigate high temperatures in urban landscapes. *Landscape and Urban Planning* **134** 127–138
[17] Gomes M, Silva C, Valadas A and Silva M 2019 Impact of vegetation, substrate and irrigation on energy performance of green roofs in Mediterranean climate *Water.* **11(10)** L2016
[18] Miller C 2016 *Extensive vegetative roofs* Whole Building Design Guide [Online]. Available: http://www.wbdg.org/resources/extensive-vegetative-roofs.
[19] Akbari H and Konopacki S 2005 Calculating energy-saving potentials of heat-island reduction strategies. *Energy Policy* **33** 721-756
[20] Rey C, Franco N, Peyre G, and Rodriguez J 2020 Green roof design with engineered extensive substrates and native species to evaluate stormwater runoff and plant establishment in a neotropical mountain climate *Sustainability.* **12(16)** L6534
[21] Cascone S, Coma J, Gagliano A and Perez G 2019 The evapotranspiration preocess in Green roofs: a review *Building and environment* **147** 337-355
[22] Alexandri E and Jones P 2008 Temperature decreases in an urban canyon due to green walls and green roofs in diverse climates. *Building and Environment.* **43** 480-493
[23] Cheng C Y, Cheung K and Chu L M 2010 Thermal performance of a vegetated cladding system on facade walls. *Building and Environment.* **45** 1779-1787
[24] Pérez G, Coma J, Martorell I and Cabeza L 2014 Vertical Greenery systems (VGS) for energy saving in buildings: *Renewable and sustainable Energy reviews.* **39** 139-165
[25] Bolund P, Hunhammar S 1999 Ecosystem services in urban area. *Ecological Economics.* **29** 293-301
[26] Gregoire B and Clausen J 2011 Effect of a modular extensive green roof on stormwater runoff and water quality. *Ecological Engineering.* **37** 963–969
[27] Karczmarczyk A, Bus A, Baryła A 2018 Phosphate leaching from green roof substrates-can green roofs pollute urban water bodies? *Water.* **10(2)** L199
[28] Charalambous K, Bruggeman A, Eliades M, Camera C and Vassiliou L 2019 Stormwater Retention and reuse at the residential plot level green roof experimental and water balance computations for long term use in cyprous *Water* **11(15)** L1055
[29] Pérez G, Coma J, Barreneche C, Gracia A, Urrestarazu M, Burés S, Cabeza L 2016 Acoustic insulation capacity of Vertical Greenery Systems for buildings *Applied Acoustics.* **110** 2018-226

[30] Van Renterghem T, Hornikx M, Forssen J and Botteldooren D 2013 The potential of building envelope greening to achieve quietness. *Building and Environment.* **61** 34-44

[31] Rowe D B. 2010 Green roofs as a means of pollution abatement. *Environmental Pollution.* **159(8-9)** 2100-2110

[32] Braaker S, Ghazoul J, Obrist M K, and Moretti M 2014 Habitat connectivity shapes urban arthropod communities: the key role of green roofs. *Ecology.* **95(4)** 1010-1021

Remotely piloted aircraft system (RPAS) use in construction: a thermography inspection case study

M Videras Rodríguez[1], A Sánchez Cordero[1], S Gómez Melgar[2, a] and J M Andújar Márquez[2]

[1]Programa de Doctorado de Ciencia y Tecnología Industrial y Ambiental, Escuela Técnica Superior de Ingeniería, Universidad de Huelva, CP 21007 Huelva, Spain
[2]TEP192 Control y Robótica, CITES, Escuela Técnica Superior de Ingeniería, Universidad de Huelva, CP 21007 Huelva, Spain
[a]sergomel@uhu.es

Abstract. The inspection of buildings with thermographic cameras is a powerful and non-invasive means of monitoring and diagnosing the state of buildings. In the field of energy efficiency, the use of thermography allows detecting insulation defects, humidity detection, location of water leaks, detection of air leaks, among others. The implementation of thermal devices in aircraft facilitates the recognition of the efficiency status of buildings and solves problems that compromise operator safety. However, it is a complex application that requires a step-by-step methodology where all the conditioning factors that can negatively affect the results are collected. This research presents a working methodology to perform energy inspections on roofs using a RPAS (Remotely piloted aircraft system) equipped with a radiometric thermal camera. In the methodology presented were detailed the premises to be taken into account before carrying out an operation of this type, the different stages related to the execution of the flight, the materials used for the acquisition of thermal images and the post-processing techniques to obtain the results of the energetic inspection. The results showed remarkable apparent temperature changes on the roof surface, in special, in the north-west part of the building. Weaknesses related to heat loss were identified in the flat roof borders and the sloping roof ridges. The execution of the inspection demonstrated the viability of using the aerial thermography for monitoring the performance of a building by identifying energy problems in advance, documenting and correcting them before they become worse and more costly to repair.

1. Introduction

The inspection of buildings with thermographic cameras is a powerful and non-invasive means of monitoring and diagnosing the state of buildings. In the field of energy efficiency, the use of thermography allows detecting insulation defects, humidity detection, location of water leaks, detection of air leaks, identification of thermal bridges, location of leaks in domestic hot water systems and air conditioning systems, detection of electrical faults, among others. The implementation of thermal devices in aircraft facilitates the recognition of the efficiency status of buildings and solves problems that compromise operator safety. The use of RPAS (Remotely piloted aircraft system) allows these tasks to be carried out quickly and accurately, reducing operational costs and minimizing security risks [1]. However, it is a complex application that requires a step-by-step methodology where all the conditioning factors that can negatively affect the results are collected.

Some authors have conducted literature reviews that address the identification of standard procedures for operating RPAS in energy audit missions [2]–[4]. Julian Moore et al. (2018) proposed a simple, cost-effective method for performing site inspections of water, heat, and ventilation systems in 30 buildings on a campus [5]. Juan Ortiz-Sanz et al. (2019) tested the use of aerial thermography to detect thermal bridges, air leaks and the presence of humidity in a semi-buried building [6].

In this research, it presents a working methodology to perform energy inspections and detect heat leakages on roofs using a RPAS equipped with a radiometric thermal camera. Heat leakage is the undesirable and typically uncontrolled loss or gain of energy through the building envelope. Heat leakage can occur through thermal bridging or excessive heat loss, air leakage and the lack of or damage to thermal insulation in building elements [7].

The aim of this experiment was double: firstly, to carry out a flight with a RPAS to identify the different tasks and processes involved in preparing and carrying out an inspection work; secondly, to apply infrared thermography to diagnose the energy status of a building built in 1962 that takes part of a future rehabilitation and maintenance project.

The development of the case study consists of different stages: the identification of the materials and software used to carry out the energy inspection, the planning of the RPAS mission, the execution of the flight, the post-processing task of the telemetry acquired and the analysis of the results obtained from the aerial thermography.

This research follows the next structure: The materials and methods used for the execution of the case study are detailed in section 2. The results acquired in the case study are shown in section 3 and they are discussed in section 4. Finally, the conclusions obtained throughout the development of this research are presented in section 5.

2. Materials and methods

The inspection was carried out in a 58-year-old building belonging to the University of Huelva, the Von Neumann building located near the former Higher Technical School of Engineering in La Rabida (37º12'9.2"N 6º55'15.8"O) (Figure 1). This university building is part of a future campus rehabilitation and maintenance project. Prior to the design of this project, an aerial inspection was carried out to detect and document possible energy problems in advance and include their correction in the project report. The aerial inspection carried out with the RPAS focused on the diagnosis and analysis of the building's roof.

Figure 1. Von Neumann building.

The characteristics of the building's horizontal envelope were identified. Several types of roof were differentiated in the building:

- Tile single-sloped roof in the four perimeter blocks.
- Tile hip-roof in the central block.

- Flat roof with different materials in the section that connects the central block with the perimeter ones. The materials identified on the flat roof were: ceramics of different colors, brick and glass (Figure 2).

Special attention was given to the materiality of the roof to be inspected by thermography because it is necessary to know its emissivity to apply this technique correctly. Thermal cameras measure the infrared radiation emitted by objects. This depends on the temperature of the body and the emissivity of the material. It does not depend on the color of the material. The color of the visual spectrum has no perceptible effect on the infrared radiation emitted by a body. It is true that colors absorb visible light, and since solar radiation is higher in the visual spectrum than in the infrared, some colors absorb more radiation than others and therefore get hotter (and radiate more to the camera), but it does not depend on the emissivity [8].

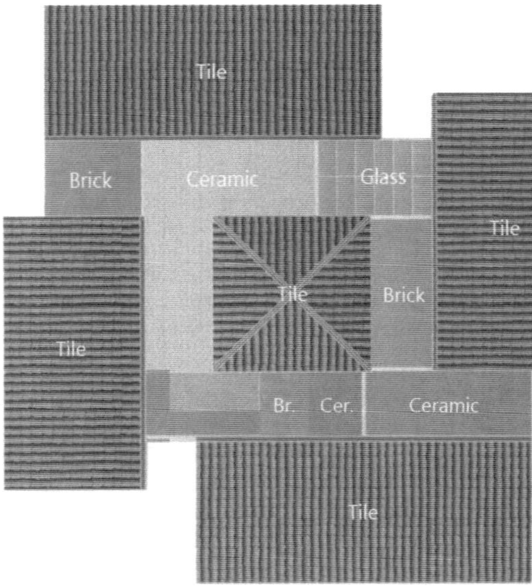

Figure 2. Materials identified on the roof of the building.

It should be noted that if the emissivity of the material is less than 0.60, an accurate temperature reading will not be obtained using infrared technology. Thermography is especially useful for materials with emissivity higher than 0.8. In these cases, the apparent body temperature is close to the real temperature. The emissivity values of the materials identified were checked using standard emissivity tables and it was confirmed that all of them had an emissivity in the range of 0.92- 0.94 (Table 1). However, glass also has high reflectivity and reflects ambient radiation rather than emitting its own, giving inaccurate temperature readings. For this reason, this area was not taken into account in the analysis of the inspection results.

Table 1. Emissivity and roofing material.

Material	Emissivity (ε)
Tile	0.93
Brick	0.94
Ceramic	0.92
Glass	0.93

2.1. Materials and software tools

This section shows the materials and software used to carry out the energy inspection. The first step in flight execution is mission planning. For this, the software DJI Flight Planner™ (DJI, Nanshan, Shenzhen, China) was used [9]. For the execution of the flight, the DJI Matrice 100 quadcopter (DJI, Nanshan, Shenzhen, China) equipped with a radiometric thermal imaging camera FLIR Vue Pro R 640 (FLIR, Wilsonville, Oregon, USA) was employed [10], [11]. The flight operation was performed using the Litchi™ application and the camera parameter settings in the FLIR™ software [12], [13]. To use both, a smartphone device was connected to the RPAS's transmitter. The post-processing phase of the collected aerial images was carried out by the Pix4Dmapper™ software (Pix4D, Lausanne, Switzerland) [14].

2.2. Planning of the RPAS mission

The planning of a flight in a wide and spacious inspection area can be easily executed by generating simple polygons, regular squares or circular trajectories using specialized software. It is necessary to perform a preview inspection to know the heights of the scene and to design a flight plan without collisions. The task becomes more complex if the RPAS's trajectory is carried out in densely built urban areas where the surrounding environment must be considered and safety distance conditions maintained from any obstacle. This type of mission also requires manual operation by experienced pilots during the flight to ensure safety [15].

The most common method for automated aerial imaging is to use a standard flight planner such as DJI Flight Planner™ [9], PrecisionHawk™ [16] or PixHawk Ardu Planner™ [17]. In this experiment, the mission planning was carried out by the DJI Flight Planner™ software, which allows the drawing, based on a satellite image, of the waypoints that the aircraft will follow during the flight (Figure 3).

Figure 3. Scheme of the path planning.

The entire flight trajectory was automated as the building that was inspected is located far from the urban center. The heights of the elements surrounding the building were secured to avoid any type of collision and to establish an appropriate flight height. Using the software, a square shaped polygon was drawn on the roof of the building to be inspected and data such as: flight altitude above ground level (18 meters), air speed (5 km/h) and image overlap (85%) were set to automatically estimate the number of photos needed to cover the study area. With these specifications, the average flight time established was approximately 30 minutes, requiring two batteries.

2.3. Execution of the flight

In order to carry out thermography correctly, special attention was paid to several factors like the time of the flight inspection which was carried out early in the morning to avoid possible reading errors due to direct sun incidence.
Weather conditions were also key: a day with light cloud cover and little wind was chosen. It was ensured that there was no rainfall in the previous days to avoid surfaces wet from rain or exposed to it for several hours as the water reduces its temperature considerably. In addition, a thermal difference of more than 10ºC was guaranteed between the interior and exterior of the building, to facilitate the identification of insulation defects in the building's roof. The outside temperature was measured at 14ºC. The heating systems in the building's labs were set at 26ºC from the day before the inspection until the end of the flight. However, this indoor temperature was not assured in all rooms of the building due to the lack of these devices in all areas. Even so, the transfer of hot air from the laboratories to the other rooms was facilitated by opening doors prior to inspection. This limitation was taken into account in the interpretation of the results. Two flight tests were carried out prior to the one shown in this research to define and optimize the whole process carried out during the case study.
The flight was made early in the morning on the 19th of November 2020, at 08:42 am with a duration of 28 minutes in La Rabida, Huelva (UTC+1). The DJI Matrice 100 quadcopter was used, which is a fully programmable multirotor platform that can be customized with different sensors for specific applications (Figure 4). This RPAS is powered by four rotors that allow its vertical take-off and landing. The dual battery components allow for a flight range of up to 40 minutes and its rigid and lightweight structure allows for the incorporation of up to 1 kg of payload. It can reach a maximum cruising speed of 22 m/s and support the wind resistance of up to 10m/s. It is controlled by a transmitter that allows a maximum transmission distance of 5 km on an operating frequency between 5725 and 5825 GHz [10].
The payload included a radiometric thermal camera FLIR Vue Pro R 640 in order to gather thermal infrared measurements. This sensor is sensitive to the mid-infrared spectral band between 7.5-13.5 μm. The resolution of the camera is 640x512 pixels equipped with a 45 degree lens providing 327,680 measuring points in the image. It has an accuracy of (+/-) 5ºC and a thermal sensitivity of 0.05ºC. The FLIR camera is connected to the RPAS by a single USB cable and does not require a sensor control module [11].
The mission planning was exported directly to the Litchi™ software to perform the flight [12]. This application allows the RPAS to take off, fly the mission and land safely after automatically capturing all the necessary images. Tracking through the waypoints was ensured in real time with this application (Figure 5). The radiometric thermal camera equipped on the RPAS was connected via Bluetooth to the FLIR™ software to set the camera settings and data capture [13]. The data collected did not record the GPS locations of each image during collection. The camera settings set were: range of humidity (>60%), air temperature (14ºC) and emissivity of the materials to be inspected (0.93).

Figure 4. DJI Matrice 100 equipped with the thermal camera FLIR Vue Por R 640.

Figure 5. Litchi™ software interface during the mission.

2.4. Post-processing of aerial images

Once the flight was finished, the 976 images captured were processed by the Pix4Dmapper™ [14]. Pix4Dmapper™ is commonly used for photogrammetry and professional mapping using RPAS. This software transforms images and allows to create and edit point clouds, digital surface models (DSMs), meshes and orthophotos. These processes are performed automatically due to its processor core based on photogrammetry techniques and artificial vision algorithms [18]. By means of a grid interpolation of the dense point cloud obtained by the sensor, a three-dimensional model of the building was obtained. All the individual images were combined into an orthomosaic thermal image to obtain an orthophoto of the entire area of interest.

3. Results

This section presents the results obtained from the thermal inspection carried out by RPAS in the case study. A qualitative analysis was carried out during this inspection. In this type of analysis, the thermal image is analyzed to reveal anomalies of varying magnitude. These are located and evaluated, but not quantified, as the work is carried out at apparent temperatures. When thermography is used qualitatively, the temperature scale is not useful. Quantitative analyses are rarely carried out, especially in interior rooms, because certain specific conditions are required and data must be obtained in situ on the real values of the material's emissivity and the apparent temperature reflected. Differences in emissivity between materials complicate quantitative inspections that require precise temperature measurements [8]. For this reason, many specialists choose to carry out qualitative inspections to focus on the temperature difference such as follows in this section.

Once the flight was completed and the thermal images acquired were processed, a point cloud grid and a 3D model of the whole building was obtained using the Pix4Dmapper™ software (Figure 6).

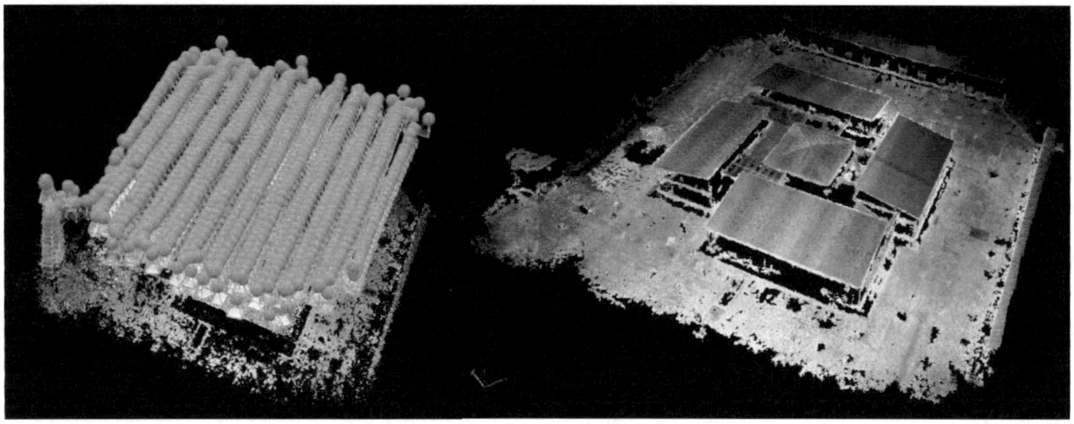

Figure 6. Point cloud grid and 3D building model.

All the individual images were combined into an orthomosaic thermal image to obtain an orthophoto of the entire area of interest (Figure 7). It is shown next to a figure which includes the ground floor and the roof floor of the building to facilitate the identification of insulation defects.

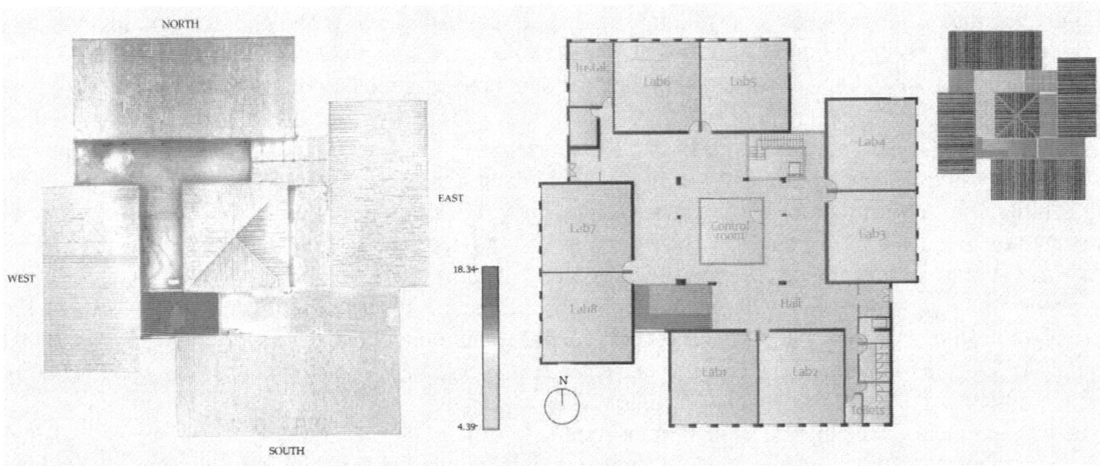

Figure 7. Thermal ortophoto and building planimetry.

The thermal image shows higher thermal radiance in the northern perimeter bock (containing laboratories 5 and 6 and the installation room) than the rest of the perimeter blocks. In this same block, there is also a considerable difference in apparent temperature in the roof over the laboratory room and the installation room. The west and est perimeter blocks show similar thermal radiance patterns and there is a slight increase of apparent temperature in the upper areas near the ridge roof. The southern block shows a different color pattern between the eastern and western part, with more thermal radiance in the western part over lab 1. The central block shows a high apparent temperature in the part of the ridge roof facing south-east orientation.

The flat roof shows significant thermal radiance differences in a single room that functions as a lobby. The northwestern area covered with bricks and ceramics shows large apparent temperature exchanges in the whole surface and at the borders of the roof that are in contact with the walls that form the slope of the inclined roof. However, the south-east area covered with the same materials, and especially the access area to the east façade and the toilets, is the one that shows the lowest apparent temperatures. In this part of the roof there is a slight increase of apparent temperature in the borders.

As specified in materials and methods section, the part of the window cover was not analyzed due to the high reflectivity of the material that reflects ambient radiation rather than emitting its own, giving inaccurate temperature readings.

4. Discussion

This section discussed the results obtained in the section 3. The changes in apparent temperature identified can be related to problems with the energy performance of the building:

- The north block shows higher surface apparent temperatures than the others, which is evidence of poor energy performance due to heat losses in the roof of this area of the building. The difference in roof thermal radiance between the installation and laboratory rooms may be due to: i) the difference in internal temperatures (there is no HVAC system in the installation room to ensure the ten-degree difference with the outside), ii) poorer thermal insulation in the laboratory area.
- The east and west blocks show slight increase of apparent temperature in the upper areas near the ridge roof where normally heat is accumulated due to the rise of hot air under the roof.
- The central block shows a high apparent temperature in the part of the ridge facing south-east orientation, although this is possibly due to the slight incidence of sunlight at the time of inspection.
- The flat roof shows poor energy efficiency in the northwest area. The whole flat roof shows variations in apparent temperature between the different surfaces although no radiance pattern is repeated between the same materials. The defects may be due to the execution of the roof because heat leaks are detected in the borders with walls.

These problems could be solved by adding continuous insulation panels on all roofs. In the case of sloping roofs, the ridge areas should be covered to prevent heat leakage. And in the case of flat roofs, by covering the areas between different materials and their borders in contact with the vertical wall that forms the sloping roof.

Below, they are a series of recommendations and analyses carried out by the scientific community that can complement this research in future works before defining a definitive and conclusive solution.
For example, it would be useful to carry out further inspections using aerial techniques at a lower height and with a different angle of vision to obtain more precise images [19].
This task can also be complemented by the use of a variety of sensors and manual thermal cameras inside the building. Other systems such as the blower door test (EN 13829) is useful to detect infiltration/exfiltration regions and thermal bridges. Burak Kakillioglu et al. (2018) applied aerial and manual techniques of infrared technology to study the thermal performance of buildings. They concluded that ground-based thermography complemented aerial techniques by providing more accurate identification of heat loss in buildings [7]. Mohamed Larbi et al. (2020) complemented the use of infrared cameras with other devices such as thermocouples, fluxmeter and pressure and humidity sensors to diagnose the energy performance of a building. The experiment was carried out by collecting measurements every second for more than a week [20].
The measurement of the thermal transmittance value U of the building envelope can also be measured by thermography and complements the energy audit. This coefficient takes into account all the heat transfer processes that are taking place through a building envelope. The precise evaluation of energy leaks can be quantified by knowing the value of thermal transmittance [21]. However, the difference between theoretical and actual thermal transmittance obtained by infrared techniques can be considerable and can lead to an over or underestimation of energy efficiency. For this, it would be necessary to carry out a quantitative thermography that provides real temperature values [22], [23].
This energy audit can be complemented by using aerial thermography to diagnose other problems in the building related to energy efficiency such as mold growth and the presence of water condensation between the walls and the roof as carried out by David Glew et al. (2017) [24]. Wet parts of a roof lose heat more quickly than dry areas, as moisture is a better conductor of heat energy than air. Thermography allows to detect heat and to distinguish this pattern of irregular heat dissipation.
The efficiency of thermal insulation in buildings has a direct impact on energy consumption for heating and ventilation. Therefore it will also be necessary to compare the results obtained in thermography with the electricity consumption data of the building. As well as checking the HVAC equipment to supervise its correct operation which can seriously affect the results obtained through thermography.
To achieve substantial energy savings by improving existing buildings is the result of accurate and reliable energy audits, followed by adaptation strategies that respond accordingly. An inaccurate energy audit may result in lower than expected energy savings, no energy savings or even an occasional increase in energy use [7]. For this reason, the energy audit shown in this case study should be complemented with other techniques derived from the literature review, in order to avoid a worse environmental impact and a waste of investment capital.

5. Conclusions

An experiment in aerial thermography was carried out in a 58-year-old building, whose progress was presented step by step in order to facilitate its understanding and offer the possibility for other researchers to carry out similar maintenance and inspection tasks. The materials and software used to carry out the energy inspection were defined. The mission planning process was also detailed, as well as the execution of the flight with the different requirements to be taken into account to perform a correct thermography and the post-processing of the 976 images obtained to finally generate a three-dimensional model of the building and a single thermal orthophoto.
The application of thermography in the building made it possible to diagnose heat leakages in the roof, documenting them in advance before they become worse and more costly to repair. The execution of the case study demonstrated the viability of using this non-invasive technique for monitoring the energy status of a building. Weaknesses related to heat loss were identified in the flat roof borders and the sloping roof ridges. The results showed remarkable apparent temperature changes on the roof surface, in special, in the north-west part of the building (both on the sloping roof and on the flat roof). These problems could be solved by adding continuous insulation panels on all roofs. However, before defining a conclusive solution it would be necessary to carry out further inspections that provide

accurate and reliable results and do not aggravate the energy performance of the building or lead to wasted economic expenditure.

For this reason, different recommendations for future work were proposed to complement the energy audit carried out and presented in this investigation. Some examples are: the repetition of the aerial inspection with RPAS changing the angle of vision, at a lower altitude and incorporating the view of the facades; the complementary use of other techniques such as manual thermal cameras or humidity sensors inside the building; the use of systems such as the blower door test (EN 13829) that allows the identification of thermal bridges and air leaks with greater accuracy; the calculation of the envelope's real thermal transmittance factor; the detection of other pathologies such as the presence of humidity and condensation; the maintenance and revision of air conditioning systems (HVAC) and the analysis of the building's electricity consumption in order to quantify the consequences generated by the envelope's pathologies on economic costs.

The proposed methodology, the results obtained and the recommendations suggested for future work based on scientific literature, can be extrapolated to other buildings and be the basis of knowledge for other researchers and building designers who are committed to the implementation of new technologies for the energy audit of buildings.

6. References

[1] Rakha T and Gorodetsky A 2018 Review of Unmanned Aerial System (UAS) applications in the built environment: Towards automated building inspection procedures using drones *Autom. Constr.* **93** 252–64

[2] Bayomi N, Nagpal S, Rakha T, and Fernandez J E 2021 Building envelope modeling calibration using aerial thermography *Energy Build.* **233** 110648

[3] Sadhukhan D, Peri S, Sugunaraj N, Biswas A, Selvaraj D F, Koiner K, Rosener A, Dunlevy M, Goveas N, Flynn D, and Ranganathan P 2020 Estimating surface temperature from thermal imagery of buildings for accurate thermal transmittance (U-value): A machine learning perspective *J. Build. Eng.* **32** 101637

[4] Shariq M . and Hughes B . 2020 Revolutionising building inspection techniques to meet large-scale energy demands. A review of the state-of-the-art.pdf

[5] Moore J, Tadinada H, Kirsche K, Perry J, Remen F, and Tse Z T H 2018 Facility inspection using UAVs: a case study in the University of Georgia campus *Int. J. Remote Sens.* **39** 7189–200

[6] Ortiz-Sanz J, Gil-Docampo M, Arza-García M, and Cañas-Guerrero I 2019 IR thermography from UAVs to monitor thermal anomalies in the envelopes of traditional wine cellars: Field test *Remote Sens.* **11**

[7] Kakillioglu B, Velipasalar S, and Rakha T 2018 Autonomous heat leakage detection from unmanned aerial vehicle-mounted thermal cameras *ACM Int. Conf. Proceeding Ser.*

[8] Testo SE & Co. KGa 2017 *Thermography, Pocket guide*

[9] Flight planning software for DJI drones - DJIFlightPlanner web page. https://www.djiflightplanner.com/ (accessed Nov. 10, 2020)

[10] Matrice 100 DJI web page. https://www.dji.com/es/matrice100 (accessed Nov. 10, 2020)

[11] FLIR Vue Pro R - FLIR Systems web page. https://www.flir.es/products/vue-pro-r/ (accessed Nov. 10, 2020)

[12] Litchi for DJI Mavic / Phantom / Inspire / Spark web page. https://flylitchi.com/ (accessed Nov. 10, 2020)

[13] App FLIR Tools - FLIR Systems web page. https://www.flir.es/products/flir-tools-app/ (accessed Nov. 10, 2020)

[14] Pix4Dmapper - Pix4D web page. https://www.pix4d.com/es/producto/pix4dmapper-fotogrametria-software?utm_term=&utm_campaign=Website+traffic-Search-Espanol+mapper&utm_source=adwords&utm_medium=ppc&hsa_acc=5624998817&hsa_cam=2062345625&hsa_grp=104577522673&hsa_ad=444355708942&hsa_src=g&hs (accessed Nov. 10, 2020)

[15] Koch T, Körner M, and Fraundorfer F 2019 Automatic and semantically-aware 3D UAV flight planning for image-based 3D reconstruction *Remote Sens.* **11** 1–29
[16] PrecisionHawk web page. https://www.precisionhawk.com/ (accessed Nov. 10, 2020)
[17] PixHawk Ardu Planner web page. https://ardupilot.org/planner/ (accessed Nov. 10, 2020)
[18] Pricope N G, Mapes K L, Woodward K D, Olsen S F, and Britton Baxley J 2019 Multi-sensor assessment of the effects of varying processing parameters on uas product accuracy and quality *Drones* **3** 1–17
[19] Ficapal A and Mutis I 2019 Framework for the detection, diagnosis, and evaluation of thermal bridges using infrared thermography and unmanned aerial vehicles *Buildings* **9**
[20] Larbi Youcef M H A, Feuillet V, Ibos L, and Candau Y 2020 In situ quantitative diagnosis of insulated building walls using passive infrared thermography *Quant. Infrared Thermogr. J.* **00** 1–29
[21] Nardi I, Lucchi E, de Rubeis T, and Ambrosini D 2018 Quantification of heat energy losses through the building envelope: A state-of-the-art analysis with critical and comprehensive review on infrared thermography *Build. Environ.* **146** 190–205
[22] O'Grady M, Lechowska A A, and Harte A M 2018 Application of infrared thermography technique to the thermal assessment of multiple thermal bridges and windows *Energy Build.* **168** 347–62
[23] O'Grady M, Lechowska A A, and Harte A M 2017 Quantification of heat losses through building envelope thermal bridges influenced by wind velocity using the outdoor infrared thermography technique *Appl. Energy* **208** 1038–52
[24] Glew D, Brooke-Peat M, and Gorse C 2017 Modelling insulated coving's potential to reducing thermal bridging and moisture risk in solid wall dwellings retrofitted with external wall insulation *J. Build. Eng.* **11** 216–23

Study on new test technology of pile foundation bearing capacity

Zhang Xue-feng[1]
[1]Research Institute Of Highway Ministry Of Transport, 100088, Beijing, China
28144484@qq.com

Abstract. A revolutionizing pile foundation testing technology, STATNAMIC has been analysis in the paper. This method is the testing method of bearing capacity for the pile foundation that has the advantages such as economical, speedy, accurate and reliable etc. By burning solid fuel in a combustion chamber, a controlled and pre-determined load is applied directly to the pile foundation without introducing high tension forces. The duration of loading is substantially longer than both the impulse produced by dynamic testing and the natural frequency of pile foundation. Moreover, owing to test without anchor pile, thus it would be able to save cost of manufacturing the anchor pile and the constructing period. And, finally it is not limited to the type of pile、 the pile gradient and the surroundings of pile.

1. Introduction
STATNAMIC test pile method is an international advanced pile load test method, 1989 was successfully developed by Berming Hammer and the Institute of Construction and Research (TNO) of the Royal Academy of Sciences in the Netherlands [1], [2]. Through a special device to change the impact of the dynamic test for the gentle continuous thrust, then it can be achieved through a smaller weight to a higher load level and obtain a decomposable load test curve, and finally through the analytical treatment to obtain static load curve. STATNAMIC test pile method not only applied enough pressure, but also for a certain period of time to achieve the effect of static [3], [4]. STATNAMIC test pile method is simple, relative cost lower than the static load, more importantly, the method itself is reliable and accurate, and it has been in the Netherlands, Canada, the United States, Germany, Japan, Korea and Singapore and other countries recognized and applied.

2. Methods
The pile foundation static load test method and the STATNAMIC method have their own characteristics, and their own test methods and data analysis systems have been formed. However, there are still some controversies about the scope of application of the test. Therefore, the application scope of the STATNAMIC method should be studied to clarify the applicable conditions of the STATNAMIC method.

The research on the scope of application of the STATNAMIC method is based on the collection of the existing STATNAMIC method test and theoretical analysis data, from the test principles, test data collection methods, test results reliability, and test methods. The application scope of the STATNAMIC method is studied in the aspects of economy, the destructiveness of the test process to the pile, and the dependence of the test method on external conditions.

3. STATNAMIC test mechanism
The STATNAMIC method absorbs the advantages of the dynamic test method, and simulates the static force, can measure the bearing capacity separately [5], [6]. The STATNAMIC method uses the most accurate instrumental observation system: the load is measured by a calibrated pressure ring, and the displacement is observed with a laser sensor (see Figure 1). The acceleration is measured directly

with a built-in speed meter. The force signal and displacement signal are collected by the TNO Profound Foundation Pile Diagnostic System (FPDS) pile diagnosis system (see Figure 2), and the dynamic damping caused by the movement of the pile is deduced by the finite element method [7]. Curve, and thus get the same results with the static pressure.

STATNAMIC method The test technique connects the piston with the pressure chamber to the pile head. A back pressure block connected to the cylinder is placed on the piston [8]. The solid fuel propellant ignites in the piston, producing a high pressure gas and accelerating the back pressure block. An equal reaction force gently pushes the pile into the soil. The force applied to the top of the pile and the displacement of the pile top can be accurately measured using a load cell and a laser sensor [9], [10].The data is collected by the FPDS device and the result is an image-force-displacement curve. The mobilized capacity and static load displacement status can be displayed immediately in the field.

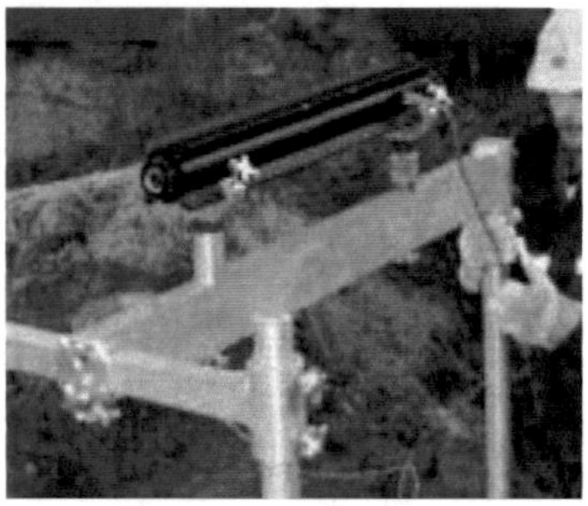

Figure1. laser sensor

Figure 2. FPDS system

The continuous phase of the Statnamic loading test is shown in Figure 3, in the figure:

a) The pre-launch phase;

b) The fuel is ignited and the combustion of the fuel produces a high pressure and accelerates the back pressure block. At this stage, the actual pile loading occurs, the reaction force pushes the pile into the earth;

c) When the fuel explodes the high pressure to hold the counteracture moment instantly, due to gravity, the surrounding loose grit quickly falls into the bottom of the press tray, forming a thick cushion layer on the pile top;

d) When the back pressure block down, the gravel to block it, the impact of gravel cushion cushioning energy consumption after the transfer to the pile around the test pile in the soil, the role of the pile at the force of only 1 %. The size and loading time of the loading during the test can be controlled by the capacity of the combustion chamber and the piston, cylinder shape, the amount of fuel, and the weight of the back pressure block.

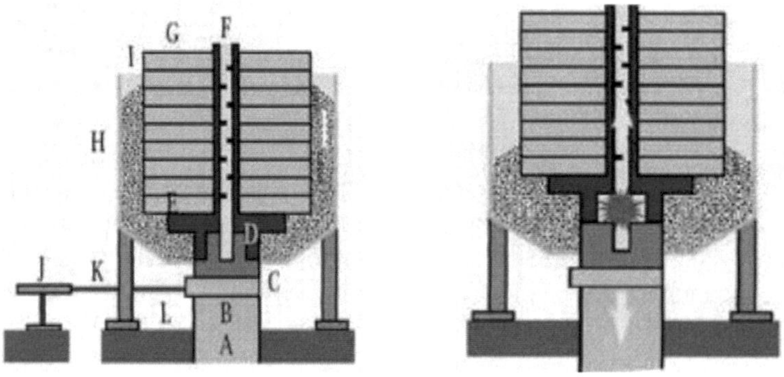

a) b)

IOP Conf. Series: Materials Science and Engineering 1148 (2021) 012005 doi:10.1088/1757-899X/1148/1/012005

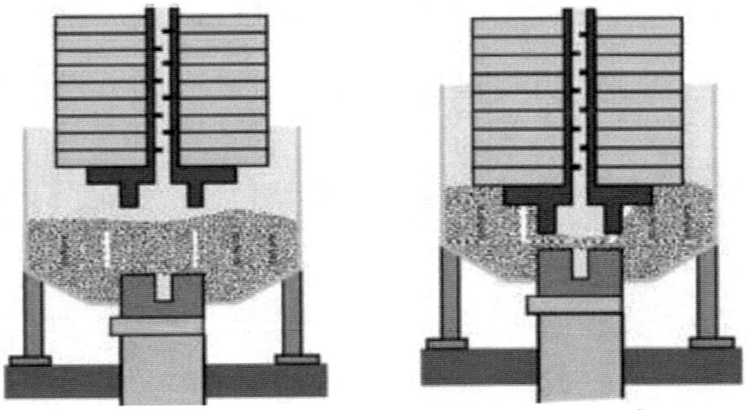

c) d)
Figure 3. STATNAMIC loading test phase diagram

In the static and dynamic state, the force distribution and displacement distribution of each particle in the pile are all the same as static, and the individual points of the pile are sinking at a certain speed, that is, the pile is sagged as a whole. At this time the force of the pile is consistent with the following equation:

$$F_{stn} = F_u + F_v + F_a \tag{1}$$

Where: F_{stn}— static and dynamic;

F_u—Static soil resistance, U is a function of displacement;

F_a—an inertial force, a function of the acceleration.

In the static and dynamic measurement, the static and dynamic load with the sensor placed in the pile head records, displacement U laser sensor with a record, acceleration a with the accelerometer records, all by the computer automatic control, ignition combustion is completed, a complete PS curve will draw The measured results are shown in Figure 4. The entire installation of the test process 2 to 3 hours to complete, the efficiency of the static load test more than 10 times.

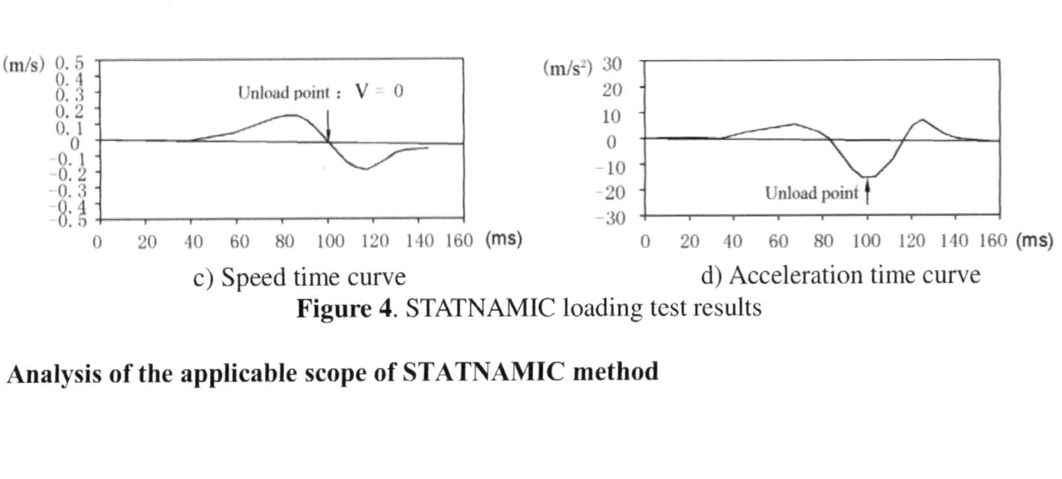

a) static motion load displacement curve b) Load and displacement time course

c) Speed time curve d) Acceleration time curve

Figure 4. STATNAMIC loading test results

4. Analysis of the applicable scope of STATNAMIC method

Through the analysis of the principle and device of static and dynamic load test (STATNAMIC), this paper summarizes the application of this method as follows:
(1) The dynamic and static load test (STATNAMIC) method can test the pile foundation in land and water environment, as shown in Figure 5 and Figure 6.

Figure 5. Terrestrial Environment Loading Test Photo

Figure 6. Deepwater Environment Loading Test Photo

(2) STATNAMIC method test technology can be used in any loading direction. Statnamic has been successfully used for ramps and horizontal loading tests, as shown in Figures 7 and 8.

Figure 7. Horizontal Loading Test Photo

Figure 8. Slope Pile Loading Test Photo

(3) At present, the tonnage of the static test method has reached 70MN. Both the steel pile, the concrete pile and the stakes, the static resistance and the dynamic resistance and the single pile and the group pile of less than 70MN can be tested with STAT-NAMIC To determine its vertical bearing capacity.

5. Result

STATNAMIC method has the following characteristics:
(1) load and displacement by the electronic load meter and laser level direct measurement;
(2) rapid test, real-time access to load ~ sedimentation curve;

(3) do not need to use anchor pile, can save a lot of anchor pile production costs and duration;
(4) is not eccentric and can cause sufficient displacement of the pile;
(5) is not limited by pile type, pile inclination and pile surrounding environment. It is especially suitable for measuring pile type of inclined pile and water working.
(6) can be carried out at the bottom of the excavation test pile.

6. Conclusion

Through the above research and analysis, we can draw the following conclusions:
(1) The STATNAMIC method is a new method that combines static and dynamic, retaining advantages and overcoming shortcomings. Compared with the dynamic measurement method, it prolongs the loading time, reduces the influence of stress wave propagation, and makes the analysis method intuitive and easy. Understand, the bearing capacity can be measured quickly and accurately;
(2) STATNAMIC method still belongs to the category of dynamic measurement method;
(3) The STATNAMIC method does not require anchor piles, is convenient and fast, greatly saves engineering costs and shortens the construction period;
(4) The STATNAMIC method solves the shortcomings of eccentricity and damage to the foundation pile caused by the heavy hammer by the high strain variation test method for large diameter bored piles, and the test principle is close to the static load test, and the final test result can be converted into the static load test result;
(5) The P-S curve obtained by the STATNAMIC method is completely measured, and various calculation basis are also obtained from the actual measurement, which overcomes the artificial arbitrariness and obtains reliable and objective results;

7. References

[1] KOVALENKO, Y., BOLOBAN, V., GONCHARENKO, I., BORYSOVA, O., MATVIEIEV, S., & MARYNYCH, I. K. V. (2020). Biomechanical assessment of statnamic stability of rhythmic gymnasts of the stage of specialized basic training. *Journal of Physical Education & Sport*, 20.

[2] Abbas, H. A., & Yasir, S. F. (2018). Comparative study of linear and non-linear analysis of Statnamic loading test for different pile lengths. doi:10.1063/1.5062639.

[3] Chew, S., Chuah, L., Tan, H., & Eng, Z. (2019, July). A Comparison Study for Static and Statnamic Load Tests on Two Instrumented Piles in Southeast Asia. *In 10th International Conference on Stress Wave Theory and Testing Methods for Deep Foundations*. ASTM International.

[4] Haque, M. N., Abu-Farsakh, M. Y., & Nickel, C. (2018). Evaluation of Design Parameters (α and β) for Open-Ended Instrumented Test Pile. *IFCEE 2018*. doi:10.1061/9780784481578.012

[5] AFT (2017). Final Report of Axial STATNAMIC Load Testing – Production Shaft P7-2 at South River Valley (SRV), Edmonton Valley Line LRT Stage 1; August 25, 2017, *Advanced Foundation Testing,* Green Cove Springs, Florida

[6] Wang, X., Tweedie, R., and Law, D. (2017). Edmonton ICE District Towers Geotechnical Investigation and Foundation Design, GeoOttawa 2017, *Canadian Geotechnical Society*, 8p.

[7] Abu-Farsakh, M. Y., Haque, M. N., & Tsai, C. (2017). A full-scale field study for performance evaluation of axially loaded large-diameter cylinder piles with pipe piles and PSC piles. Acta Geotechnica, 12(4), 753-772.

[8] Rybak, J. (2017, October). Some remarks on foundation pile testing procedures. In IOP Conference Series: Materials Science and Engineering (Vol. 245, No. 2, p. 022092). IOP Publishing.

[9] Shahbazi, M., Cerato, A. B., Allred, S., El Naggar, M. H., & Elgamal, A. (2020). Damping characteristics of full-scale grouped helical piles in dense sands subjected to small and large shaking events. Canadian Geotechnical Journal, 57(6), 801-814.

[10] M. F. Alwalan, M. H. El Naggar. (2020). Finite element analysis of helical piles subjected to axial impact loading. Computers and Geotechnics, Volume 123, July 2020, 103597.

Acknowledgments

This study was funded by Young Elite Scientists Sponsorship Program by China Association for Science and Technology. The authors thank the anonymous reviewers and the Editor for their constructive comments and advice, which greatly improved the quality of this paper.

IOP Conf. Series: Materials Science and Engineering 1148 (2021) 012006 doi:10.1088/1757-899X/1148/1/012006

Damage analysis of deep water high pile cap foundations under different concrete corrosion by ship impact load

Zhang Xuefeng
Research Institute of Highway Ministry of Transport, 100088, Beijing, China
28144484@qq.com

Abstract. The deep-water high-pile cap foundation is in the water environment for a long time, which is likely to cause corrosion damage to the pile foundation concrete. At present, the damage analysis of the upper structure has received more attention and the technology is relatively complete, but the research on the damage identification and analysis of the substructure and foundation Relatively lacking. Due to the loading characteristics of the substructure, its own bearing characteristics and the geological watermark environment, which are quite different from the upper structure, the corresponding damage identification and damage analysis evaluation technology is quite different from that of the upper structure, so special research is needed.

1. Introduction

With economic development, increasing demand for transportation and continuous improvement of bridge construction technology, a large number of river-crossing bridges at home and abroad have emerged one after another. High-pile caps have the advantages of convenient construction and saving engineering costs and are widely used by cross-river-sea bridges, which causes the deep-water high-pile cap foundation to be in a water environment for a long time. The foundation structure of deep-water high-pile caps is flexible, and it is easily damaged by the influence of foundation erosion and water medium dynamics [1], [2]. Effective damage analysis has always been a difficult point for the healthy and safe operation of the cross-river-sea bridge.

According to the available data, there are two main types of foundation diseases and damages of bridges across rivers and seas according to their causes [3]. One is durability diseases caused by environmental influences of materials, such as apparent diseases such as steel corrosion, erosion, and biological corrosion; Force damage, such as earthquake, impact, wind and wave impact, etc., causes excessive force on the structure, causing concrete cracking, local damage and internal damage. The durability diseases of concrete in water are mainly manifested in the corrosion of steel bars, and the swelling, bulging, cracking, and detachment of concrete caused by corrosion [4]. There are also concrete crisping, surface shedding and water erosion caused by freezing and thawing, water and salt erosion. Diseases such as biological corrosion [5].

With regard to concrete corrosion in marine environments, scholars at home and abroad have conducted extensive and detailed experimental and theoretical studies. As far as the research scope is concerned, it mainly includes: the corrosion mechanism of steel reinforcement in concrete in marine environment, the corrosion rate of steel reinforcement in concrete, the detection method of steel corrosion, the protection method of steel corrosion, the mechanical properties of steel after corrosion, the mechanical properties of concrete after steel corrosion [6]. The influence of corrosion on the performance of concrete structures, the resistance evaluation of reinforced concrete structures or components, and the prediction of structural durability life [7]. Through the efforts of many domestic and foreign scholars, the research in the above-mentioned fields has achieved fruitful results, and a more in-depth understanding and understanding of the corrosion of steel bars in concrete in the marine environment has been obtained. Many research results have begun to be used in actual structures and have achieved good results [8].

2. Methods

Damage can be seen as various external factors, such as corrosion, load, humidity, etc. These micro-defects continue to cause, expand and integrate. The continuous distribution of micro-defects in the solid material or internal makes the material or structure degrade in performance. The rigidity and strength of the performance decrease, a large number of cracks, even breakage and failure.

Reinforced concrete components in a seawater environment have different locations and conditions, and their corrosion mechanisms are different. Generally, we divide the marine corrosion environment into atmospheric areas, splash areas, tidal ranges, and seawater full-immersion areas. The main pier foundation damage of the cross-sea bridge mainly affects the safety performance of the foundation: concrete corrosion, concrete cracking, steel corrosion, and foundation erosion. This study uses concrete corrosion damage as an example to study structural damage indicators. Concrete corrosion is a common damage to pile foundations in seawater. The current situation of pile foundation corrosion of a coastal bridge is shown in Figure 1.

Figure 1. The corrosion status of a coastal bridge pile foundation

Seawater contains a lot of chlorides, which is highly corrosive to steel in the marine environment. Chloride ion corrosion is the main cause of steel corrosion in concrete. Concrete is composed of pores and coarse aggregate volume of cement mortar. In the concrete mixing system, in order to obtain the necessary workability, more water is added than the water-cement ratio [9]. This part of the water is retained in the concrete with free water. As the temperature decreases, the pore water begins to freeze and expand. The volume expansion is a degree of wet saturation, cement slurry penetration, the nearest pore distance, cooling rate and retention After the temperature, when the stress exceeds the tensile strength of concrete, local cracks may occur and the strength of concrete decreases [10]. After repeated freezing and thawing cycles, the damage of concrete will continue to increase and the strength will gradually disappear. In the marine environment, the concrete is completely immersed in water or under high humidity conditions. Due to the high degree of water permeability, the freezing and thawing cycle of concrete is longer than the general atmosphere due to the long-term heat caused by sea water The environment is serious. In addition, the internal cracks and surface cracks of concrete during freezing and thawing cycles are more easily penetrated by carbon dioxide and chloride ions, thereby accelerating the corrosion of steel bars.

3. Concrete damage constitutive relationship

The nonlinear mechanical behavior of concrete includes the influence of microcracks and plastic flow of these two types of microscopic mechanisms. Therefore, concrete materials have the following remarkable characteristics:

(1) The existence of regional instability accompanied by significant stiffness degradation and softening after peak strength stress;

(2) When the loading and unloading process of repeated concrete deformation exceeds a certain threshold, the unloading deformation is completely unrecoverable;

(3) The strength and ductility of confined concrete increase significantly (such as the state of compressive biaxial stress);

(4) Due to the tensile stress of the biaxial compressive strength of the lower concrete material, the tensile and compressive stress is higher than the uniaxial compressive strength, which is the so-called tensile and compression softening;

(5) Unilateral effect: There is a significant difference between compressive strength and tensile strength, especially when the wounded is pulled in a clear direction, the crack closure after the reverse tensile load will lead to the recovery of the stiffness of the concrete.

The reasonable damage constitutive relationship correctly describes the two different and irreversible processes of change. Since 1980, many scholars have begun to apply damage to the mechanical properties of concrete to study mechanics. The first job is to create specific experimental data for changes in the stress level of damage or pressure. This law is expressed by the following formula, usually called It is the damage model. At present, specific damage models have been established, mainly based on or uniaxial stress experiments, and semi-experimental and semi-empirical modeling methods are mostly adopted when establishing the model.

4. Simulation analysis

The North Channel Bridge of Hangzhou Bay Cross-sea Bridge is a double-tower, double-cable plane cable-stayed bridge. The main bridge is 908m in length, and the span combination is 70m+160m+448m+160m+70m. There are 26 bored piles with a diameter of 2.8m under each main pier, the average pile length is 125m, the bottom elevation of the pile is -125.800m, the diameter of the bored pile steel casing is φ3.1cm, the wall thickness is 20mm, and the bottom elevation of the casing is -43.500m. The cap is a hexagonal round and chamfered integral cap, the size of the cap is 48.5×23.7×6.0m, and the top elevation of the cap is +5.200m. Hangzhou bay bridge D13 pier model in Figure 2.

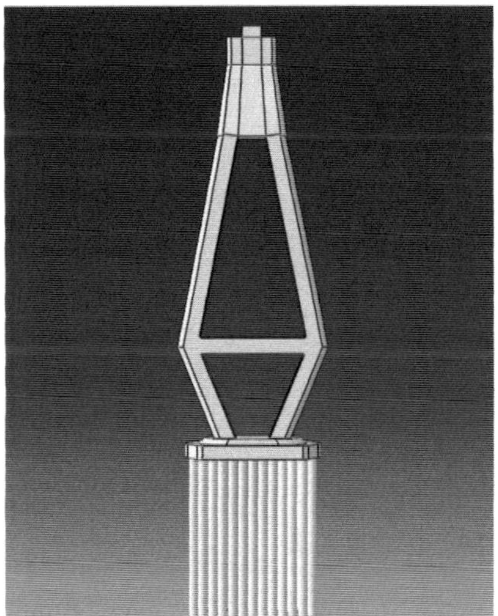

Figure 2. Hangzhou bay bridge D13 pier model

IOP Conf. Series: Materials Science and Engineering 1148 (2021) 012006 doi:10.1088/1757-899X/1148/1/012006

5. Results

In this study, the damage of the pier foundation under the load of each working condition is analyzed. The damage of the typical foundation (corrosion 15cm) is shown in Table 1, and the calculation and analysis of the bridge are shown in Figure 3 to Figure 6.

Table 1. Pier scouring case damage results

position	Concrete compression damage (DAMAGEC)	Concrete Tensile Damage (DAMAGET)	Concrete vertical stressS33(MPa)	Concret e plastic strainPE	The maximum tensile stress of steel bars (MPa)	Horizontal displacement of the tower (cm)
Cross beam in the top surface	0.00	0.00	--	0.00		
Cross bar bottom	0.00	0.00	--	-1.57E-04		
1 # Pile top outside	0.00	0.39	-18.94	0.00	162.00	6.56
1 # Pole bottom outside	0.00	0.00	2.36	0.00		
13 # Pile top outside	0.00	0.00	20.92	0.00		
13 # Pole bottom outside	0.00	0.41	-13.18	0.00		

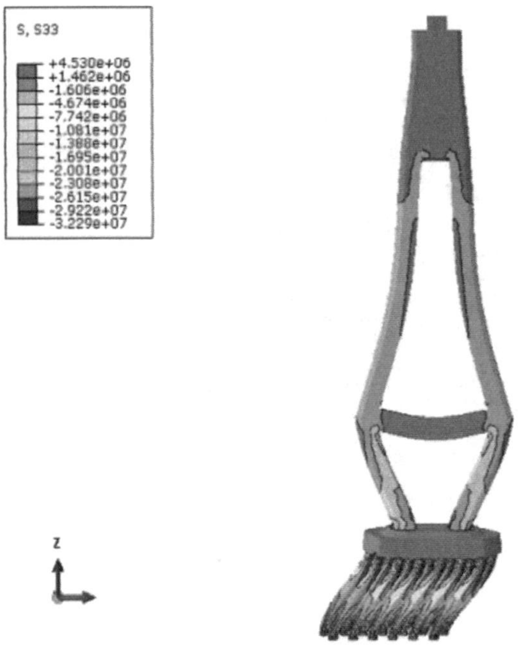

Figure 3. Pier concrete S33 strain cloud

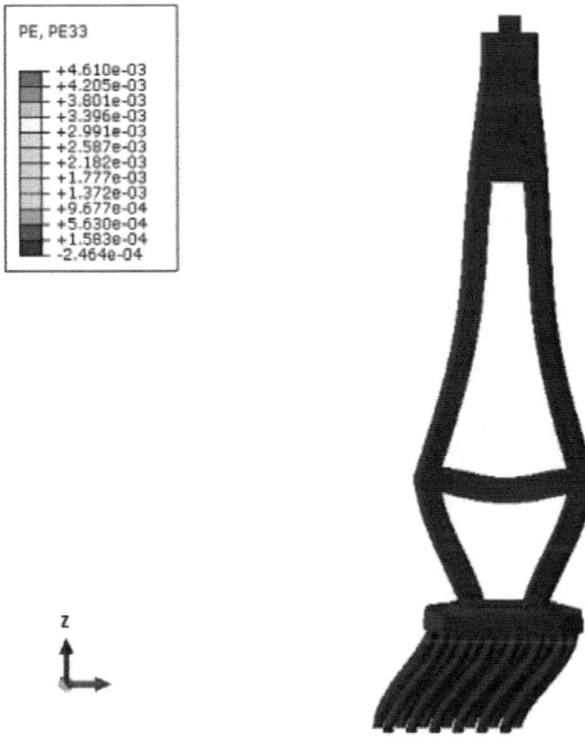

Figure 4. Pier concrete PE strain cloud

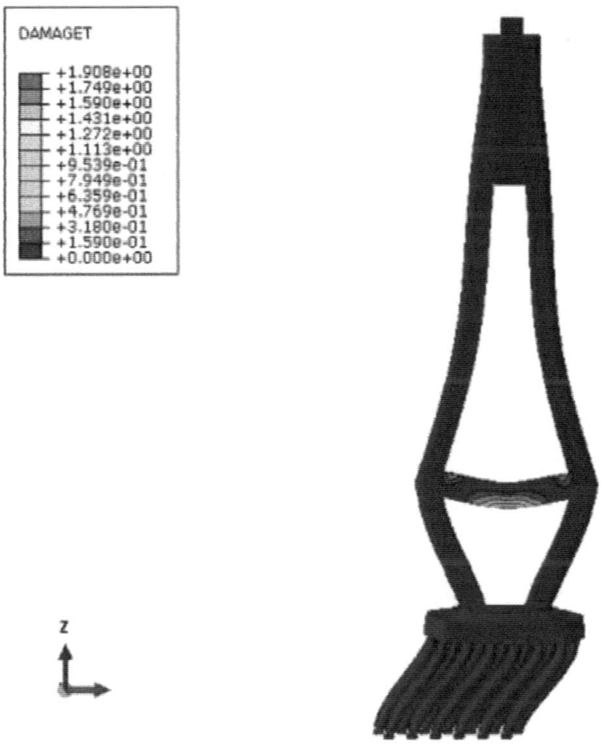

Figure 5. Main Pier DAMAGET cloud (whole idea)

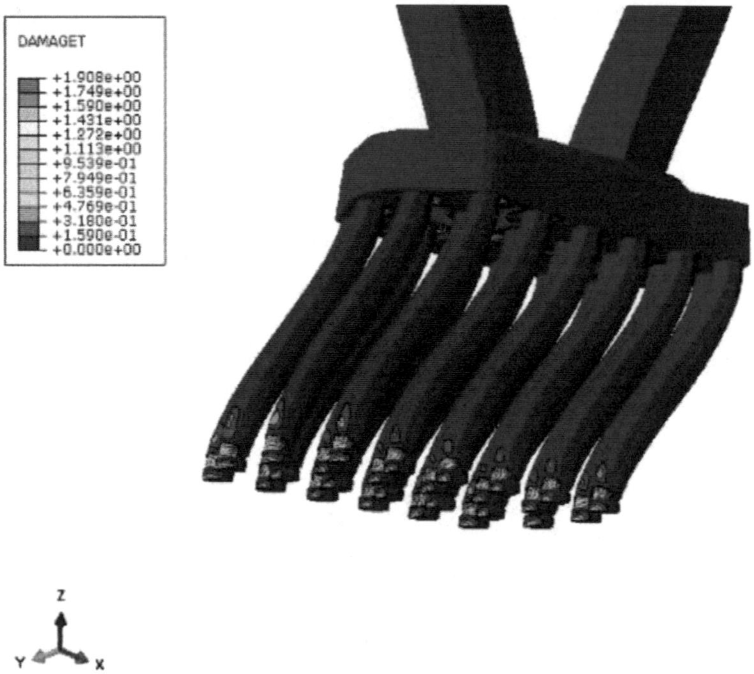

Figure 6. Main Pier DAMAGET cloud (partial view)

The analysis results show that the cracking damage range of the bridge tower beams under the ship collision is larger than that under the wave action, the top and bottom of the pile foundation cracking damage, the damage of the pile foundation corrosion damage area further expands, and the steel bar does not yield.

6. Conclusions

(1) The pier foundation of the bridge piercing 5cm under the action of the ship hit the bridge beam cracking damage, the top and bottom of the pile foundation under the local crack damage bridge pier beam damage cracking, the bottom of the pier, pile foundation and the bottom of a slight damage, Plastic strain, local cracking, rebar began to yield.

(2) Pier pier foundation corrosion in 15cm state in the ship collision load bridge beam appears more serious damage cracking, pier bottom, pile foundation at the top and bottom damage, plastic deformation, local cracking, steel yield.

(3) the pier foundation corrosion 30cm in the ship collision load bridge bridge pier serious damage cracking, the top and bottom of the pile foundation has a large area damage, plastic strain, local cracking serious, steel yield.

(4) The main displacement of the main pier foundation of the cross-sea bridge is mainly caused by the horizontal displacement of the pier top, and the buckling of the steel leads to the collapse of the structure.

7. Acknowledgments

This study was funded by Young Elite Scientists Sponsorship Program by China Association for Science and Technology. The authors thank the anonymous reviewers and the Editor for their constructive comments and advice, which greatly improved the quality of this paper.

8. References

[1] Wu, T., Qiu, W., & Wu, G. (2020). Fatigue Damage Evaluation of Pile-Supported Bridges under Stochastic Ice Loads. *Advances in Civil Engineering*, 2020.

[2] Li, C., Li, H. N., Hao, H., Bi, K., & Chen, B. (2018). Seismic fragility analyses of sea-crossing cable-stayed bridges subjected to multi-support ground motions on offshore sites. *Engineering Structures*, 165, 441-456.

[3] Xu, B., Wei, K., Qin, S., & Hong, J. (2020). Experimental study of wave loads on elevated pile cap of pile group foundation for sea-crossing bridges. *Ocean Engineering*, 197, 106896.

[4] Ti, Z., Wei, K., Qin, S., Mei, D., & Li, Y. (2018). Assessment of random wave pressure on the construction cofferdam for sea-crossing bridges under tropical cyclone. *Ocean Engineering*, 160, 335–345. doi:10.1016/j.oceaneng.2018.04.036

[5] Qinglai, F., & Yufeng, G. (2018). Effect of Reinforcement Ratio and Vertical Load Level on Lateral Capacity of Bridge Pile Foundations. *Polish Maritime Research*, 25(s3), 120–126. doi:10.2478/pomr-2018-0120

[6] Ti, Z., Zhang, M., Li, Y., & Wei, K. (2019). Numerical study on the stochastic response of a long-span sea-crossing bridge subjected to extreme nonlinear wave loads. *Engineering Structures*, 196, 109287. doi:10.1016/j.engstruct.2019.109287

[7] Xia, C., Ma, Q., Liao, Z., & Xia, H. (2018, July). Numerical Simulation of Ship Collision on RC Cap with Anti-collision Steel Box Barrier of Over-sea Bridge. *In The 28th International Ocean and Polar Engineering Conference*. International Society of Offshore and Polar Engineers.

[8] Sha, Y., Amdahl, J., Dørum, C., & Yu, Z. (2018). Numerical Investigation of the Collision Damage and Residual Strength of a Floating Bridge Girder. *Volume 11A: Honoring Symposium for Professor Carlos Guedes Soares on Marine Technology and Ocean Engineering*. doi:10.1115/omae2018-78728

[9] Aminoroayaie Yamini, O., Mousavi, S. H., Kavianpour, M. R., & Movahedi, A. (2018). Numerical modeling of sediment scouring phenomenon around the offshore wind turbine pile in marine environment. *Environmental Earth Sciences*, 77(23). doi:10.1007/s12665-018-7967-4

[10] Zhu, J., & Zhang, W. (2017). Numerical Simulation of Wind and Wave Fields for Coastal Slender Bridges. *Journal of Bridge Engineering*, 22(3), 04016125. doi:10.1061/(asce)be.1943-5592.0001002

Structural efficiency of the 'Pombalino Frontal' infilled with mortar

António Morais[1,a], Carmo Fialho[2], Basto Diogo[3] and Svetlana Ivanova[4]

[1, 2, 3, 4] Universidade de Lisboa, Lisbon School of Architecture, CIAUD, Lisbon, Portugal
[a]Abdiogo@hotmail.com

Abstract. On November 1st, 1755, occurred a huge earthquake in Portugal which hit and destroyed Lisbon. In a previous paper, where we presented some research about pombalino system, we referred and noted that we were just in 1755, when the creation and development of the 'pombalino building' occurred, so the concept of stress and deformation had not yet appeared, but, despite this, the military engineers of the kingdom were able to create a remarkable construction system with anti-seismic characteristics. In the vertical frontal truss of the Pombaline System, the spaces between the vertical and horizontal plumbs are filled with mortar that contained small elements of stone and ceramic residues. This form of filling was not carried out in order to close the space, because at the time, even in the Pombaline building, the solution adopted to close space was different. This filling was designed by the engineers to provide the building with an additional energy dissipation system, although at the time humanity has not yet developed the concepts and theory of structural analysis and, obviously, the characterization and definition of the dynamic behaviour of a structure had not yet been achieved. The frontal element design suggests their engineers' designers had at that time an accurate knowledge about the dynamic phenomenology involved in the seismic behaviour of the pombaline building. In this article we present structural calculations performed on a 'frontal pombalino' standard, used in the 'Pombalino cage', which demonstrate that the military engineers had not only a perfect understanding of the St. André cross potential, but they also know the role played by the filling mortar in the seismic response of the Pombaline building. The infill cracks when a seismic occurs and a large number of slits are formed, and each crack acts as an energy dumper (a power sink).

1. Introduction

Ordered by the prime-minister Marquis of Pombal, the military engineers of the kingdom designed the pombaline system with the main purpose of the pombaline building having an earthquake-resistant capacity. The design was done with the main intention of having the capacity to resist an earthquake, as violent as it was. Its design and creation was made with the main objective of being an anti-seismic structural system. And this purpose has been achieved. [1] As noted before, the insertion of the vertical frontal element in the 'Pombalino cage', made up of Santo André Crosse's lattices, impresses the intellectual boldness, at that time, which demonstrated the inventive genius and the remarkable knowledge revealed by the military engineers in charge of the project, concerning the effectiveness and efficiency of the structural performance achieved with the 'Pombalino building'. [2], [3] But the surprise didn't stop here. Those engineers had another powerful idea that revealed a deep understanding of the seismic behaviour and performance of the system they created and designed. They had the idea of filling the spaces between the vertical plumbs and the inclined and horizontal elements of the crosses of St. André on the vertical frontal lattice, with a lime mortar impregnated with small ceramic and stone residues. For the time, this idea was surprising and incredible. Note that the concept of stress and deformation had not yet appeared, and, therefore, mankind did not yet discover the structural calculus theory, so the concepts of bending moment and normal effort were not known at that time, but, despite this, the architects of the kingdom were able to create a remarkable construction system with anti-seismic characteristics. Those engineers knew very well that filling is a key that endows the vertical frontal lattice with extra capacity to dissipate the seismic energy.

This form of filling was not carried out only in order to close the space because, at the time, the solution adopted to close space was different. This filling was designed by the military engineers, who created the Pombaline system, to provide the building with an additional seismic energy dissipation system, although at the time the humanity has not yet developed the concepts and theory of structural analysis and, obviously, the characterization and definition of the dynamic behaviour of a structure had not yet been achieved. The mortar filling, containing small ceramic and stone residues, cracked when the earthquake oscillations started and these cracks are in a large number. Since the seismic action is oscillatory, then, each crack performs a cyclic opening movement followed immediately by the crack closing. This continuous opening/closing of the mortar cracks facilitates the dissipation of the seismic energy transmitted by the earthquake. In each oscillation cycle, each crack dissipates a certain amount of energy transmitted by the earthquake. As the cracks are high in number, so the amount seismic energy dissipated by this mechanism for opening/closing reaches a high value, relieving the structural system, flattening the maximums stresses values, which increases the seismic resistance of the Pombaline building. This mechanism actually corresponds to a tool that provides the Pombaline structural system with a behaviour similar to that of the plastic phase of contemporary materials, steel and reinforced concrete. As we know, the plastic phase acts as a fuse that prevents the increase on stresses temporarily. Perceive in 1755 how this mechanism of the mortar acts in the Pombaline seismic response is a remarkable cognitive and conceptual feat. Amazing and surprisingly how these engineers in that time could innovated and achieve this solution that performs as a plastic phase on building response, when an earthquake occurs. [4]

In this article we present the conclusions of the calculations that prove the beneficial effect of the presence of the mortar to fill the spaces between the vertical plumbs and the horizontal and inclined bars of the vertical frontal lattice with their St. André crosses.

2. Structural study of the pombalino frontal's design

2.1. Frontal design

The structural system of the pombaline building usually called pombaline cage contains in its constitution a vertical element in the entire height of the building, called Frontal that establishes the connection between the facade walls parallel to each other. The function of this vertical lattice frontal is very important, given that by establishing the connection between the facade walls parallel to each other, it forces them to move in the same direction and synchronously. The military engineers realized with the observation of the collapse of many buildings that in many buildings the facade walls remained upright and only the pavements and the roof collapsed. It was thus necessary to avoid the out-of-sync movement of the facade walls and from there they innovated and created the Pombaline frontal.

2.2. Characterization of seismic action in the frontal

The research methodology adopted to analyse the effects of the seismic action on the frontal consisted of an equivalent static analysis. For the scope and objective of the investigation, this approximation is reasonable because it is a regular structure in plan and height, with predictable dynamic behaviour. Under these conditions, and because the EC8 [5] allowed it, the effects of the seismic action were determined by applying to the structure a set of horizontal forces F_i, acting in parallel with the seismic action adopted and in the centre of the respective masses, which in this study were concentrated at the level of the respective floors.

According to EC8 5.3.2, part 1.1, its values are given by the expression $F_b = S_d(T_1) m \lambda$

Where:

$S_d(T_1)$ - ordered from the calculation spectrum for period T1;

T_1 - period of fundamental vibration for the movement considered;
m - mass of the building, corresponding to the area of influence of the front;
λ - correction factor ($\lambda = 0.85$).
In order to characterize the seismic action, the maximum acceleration of the terrain ($a_g = \gamma_1 a_{gR}$) and taking into account the values for the maximum reference acceleration in the various seismic zones in mainland Portugal, if there is a Type 2 seismic action in a 2.3 seismic zone (Lower Pombalino), we will have $a_{gR} = 1.7$ ms^{-2}. It was considered the importance factor corresponding to current buildings, importance II $\rightarrow \gamma_I = 1$. Thus, $a_g = 1.7$ ms^{-2}.
The calculation spectrum S_d (T_1), for the study conditions, considering the values of $T_B = 0.10$ and $T_C = 0.3$, obtained in EC8 (Table 3.3) $T_B \leq T \leq T_C$, is defined by the following expression:

$$S_d(T) = a_g . S . \frac{2,5}{q} . \left[\frac{T_C}{T}\right] \quad \text{(EC8 3.2.2.5)}$$

Where:
a_g – calculation value of the acceleration at the ground surface ($a_g = 1,7$ ms^{-2});
S – soil coefficient (S = 1.8);
q – coefficient of behavior (q = 2,5) EC8 (table 8.2);
T_C – upper limit of the period at constant spectral acceleration plateau;
T – vibration period of a linear system with a degree of freedom

$$T = C_t \ . H^{\frac{3}{4}} \quad \text{(EC8 4.3.3.2.2)}$$

C_t – 0.050 according to EC8 4.3.3.2.2 (3);
H – front height.

2.3. *'Frontal panel' geometric characterization*

In this study, the frontal solution is adopted, consisting of a wooden structure, with panels of 1.0 x 1.0 m2 locked by diagonals. This solution is usually presented in the characterization of the "frontal pombaline" [3]. For the overall dimensions of the 'frontal wall', a framework with 12 x 6 m2 was admitted, characterizing a building of 4 floors including the ground floor. This structure consists of plumbs, beams, "travessanhos" and diagonals, as shown in the schematic drawing presented and which is considered to be repeated four times, one by each flat, in the analyzed model.

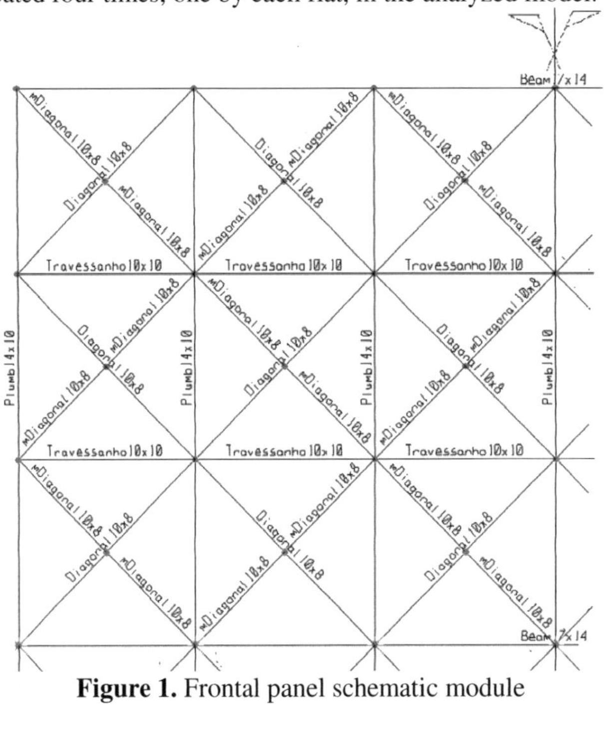

Figure 1. Frontal panel schematic module

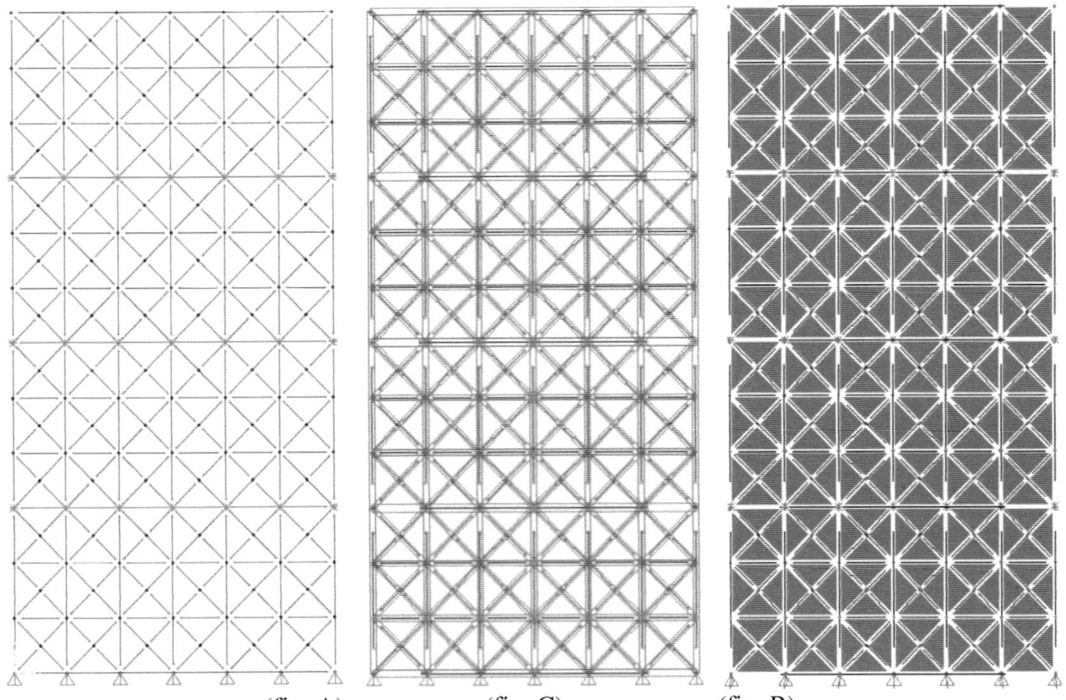

Figure 2. Frontal panel

In this paper we compare the solution of frontal wall structure without mortar infill and the solution of the framework braced by filling the panels with mortar infill. The first figure A reproduces the frontal model adopted in calculations. The other two models are identical in the formal geometric plane, adopting both locking of the panels with crosses of "Santo André". The difference between the latter two models lies in the infill of the orthogonal mesh of the lattice structures, consisting of diagonal locking posts and crossbars in each of the panels. In first model, there are no mortar infill (fig. B), the sections of all bars being identical to that of the standard structure of the 'Pombalino building'. In the other model (fig C), identical spacing of 1.0 meter were used between uprights and crosspieces but the panels are filled with mortar. This last model corresponds to the standard structure of the 'Pombalino building', according to the ones made in several works of 'Baixa Pombalina' [3], [6].

2.4. Frontal inertia forces

For the determination of the inertial forces acting on the frontal wall and for the EC8 [5], the model that substitutes the deformed of the fundamental mode by a linear configuration, considering the equivalent forces, at the level of the floor girder (Fi) resulting from the weighting of the basal force (Fb) by the parameter hi x mi, which results respectively from the height and mass corresponding to the level of the force i, obtaining for each of the floors the equivalent force given by the following expression:

$$F_i = F_b \times \frac{h_i \times m_i}{\sum_1^n h_i \times m_i} \qquad \text{(EC8 4.3.3.2.3)}$$

c/ $F_b = S_d(T_1) m\, \lambda$

In the determination of the shear force at the base, we considered the masses of the frontal wall, of the houses and overloads, corresponding to the area of influence of the frontal. In this analysis, the mass values of the elements were obtained from the usual actions for dimensioning structures, according to the following assumptions:

2.4.1. Frontal wall wood structure (c = 6m; h = 3m)

It was considered that the frontal wall structures in analysis consisted of profiles of pine wood totaling 1 m^3 per floor.

$W_{fw\,G}$=1 m^3×6 kNm^{-3}=6kN

2.4.2. Frontal wall mortar infill

It was considered that the frontal wall with mortar infill in analysis consists of triangular masonry panels with 0.15 m thick that fills the wooden structure panels of the frontal wall.

$$W_{fm\,G} = (6m \times 3m \times 0{,}15m - 1m^3) \times 20\ kNm^{-3} = 34kN$$

2.4.3. Flooring

Assuming an area of influence of 6 x 5m^2 and considering the own weight of the wood structure 0,5 kNm^{-2} and the overload 3 kNm^{-2} we will have that, at the level of each floor, the force resulting from the weight of the floor and its overload will be:

$$W_{pav\,G+Q} = 6 \times 5 \times (0{,}5 + 3) = 105\ kN$$

So, considering a), b) and c), the active vertical forces, the basal force (F_b) is obtained.

$$\Sigma W_i = 6kN{+}34kN + 105kN = 145\ kN$$

$$\therefore F_b = 4 \times 145kN = 580\ kN$$

From chapter 3.2 we obtain spectrum calculus $S_d(T)$

$$S_d(T) = a_g.S.\frac{2{,}5}{q}.\left[\frac{T_C}{T}\right] = 1{,}7 \times 1{,}8 \times \frac{2{,}5}{2{,}5} \times \left[\frac{0{,}3}{0{,}32}\right] \approx 2{,}8$$

Considering the previous components, it results that the cutting force in the base, according to the gravitational forces considered, will be:

$$F_b = S_d(T_1)m\,\lambda = 2{,}8 \times \frac{F_g}{g} \times 0{,}85 = 0{,}24F_g$$

$$F_i = F_b \times \frac{h_i \times m_i}{\Sigma_1^n h_i \times m_i}$$

For the two structural models considered the results of the inertia forces Fi in the different floors are indicated in the following tables.

The forces of inertia

Floor	h_i (m)	W_i (kN)	W_i x h_i (kNm)	F_i (kN)
4	12	111	1332	42,6
3	9	111	999	32
2	6	111	666	21,3
1	3	111	333	10,86
R/C	0	-	-	-
Σ	-	444	3830	106,6

- Frontal wall wood structure without brickwork infill:

The cutting force in the base F_b

$$F_b = 0{,}24F_g = 0{,}24 \times 444 = 106{,}56 \text{ kN}$$

- Frontal wall wood structure with brickwork infill:

The cutting force in the base F_b

$$F_b = 0{,}24F_g = 0{,}24 \times 580 = 139 \text{ kN}$$

The forces of inertia

Floor	h_i (m)	W_i (kN)	W_i x h_i (kNm)	F_i (kN)
4	12	145	1740	55,6
3	9	145	1305	41,7
2	6	145	870	27,8
1	3	145	435	13,9
R/C	0	-	-	-
Σ	-	580	4350	139

2.5. Characteristics of materials
We consider for wood (forces considered // to fibers) the acceptable values:
$\sigma_{CMax} = 25\text{MPa}$; $\sigma_{TMax} = 25MPa$; $\nu = 0{,}4$; $\gamma = 6\ kNm^{-3}$; $E = 10GPa$
For mortar infill the following characteristics were adopted:
$\sigma_{CMax} = 5MPa$; $\sigma_{TMax} = 0{,}5MPa$; $\nu = 0{,}2$; $\gamma = 20\ kNm^{-3}$; $E = 10GPa$; $\alpha_{\text{atrito}} = 11°$

3. Results

3.1. Horizontal and vertical displacement at the top of the frontal
Based on the results obtained in the structural analysis of the two models considered in this investigation, described above, the displacements, the maximum stresses and the tensions in the wooden frontal bars were calculated. The calculated maximum lateral and vertical elastic displacements are observed at the ends of the frontal end, as expected:

- Frontal wall, with section 12 x 6 m^2 with diagonals and reticulated mesh of 1.0 x 1.0 m^2. In this model, maximum lateral and vertical elastic displacements are

$U_x = 0.009$ m, $U_z = -0.006$ m;

- Frontal wall, with section 12 x 6 m^2 with diagonals and reticulated mesh of 1.0 x 1.0 m^2, considering panels with mortar infill. Masonry shear walls are modeled with appropriate material properties, previously described in 3.5, considering stiffness modifiers. The stiffness modifiers of membranes are used to simulate the behavior of cracked stage of mortar infill in the wood frame caused by seismic action. The stiffness adopted in these structural elements, to reflect a reduction in strength of the material, are reduced by membrane modifiers f_{11}, f_{22} and $f_{12} = 0{,}25$ [6]. So, with this model, maximum lateral and vertical elastic displacements are reduced to:

$U_x = 0.006$ m, $U_z = -0.003$ m;

With regard to maximum lateral displacements at the top of the frontal wall, the elastic analysis shows a reduction of U_x and U_y in the order of 1/3 and 1/2 respectively.

3.2. Efforts and tensions in frontal frame elements

In the case of the bars with bigger efforts, considered in tables 1 and 2, it can be verified that the inclusion of mortar infill in frontal wall ensures significant reductions in the tensions of the constituent elements of the wood structural elements, as shown in table 3

Table 1. Calculations without mortar infill

MAXIMUM LOADS IN FRONTAL BARS WITHOUT MORTAR INFILL							
Elements	Section (cm^2)	Bar	N_{max} (kN)	V_{max} (kN)	M_{max}(kNm)	σ(MPa)	τ (MPa)
Beam	17x14	596	59,27	11,4	- 2,01	5,5	0,72
Plumb	14x10	627	- 138,8	0,66	- 0,41	- 9,86	0,07
«*Travessanho*»	10x10	675	43,21	0,21	-0,11	4,97	0,03
Diagonal	10X8	821	- 101,03	0,87	-0,22	- 14,24	0,36

Table 2. Calculations with mortar infill considering stiffness modifiers

MAXIMUM LOADS IN FRONTAL BARS WITH MORTAR INFILL							
Elements	Section (cm^2)	Bar	N_{max} (kN)	V_{max} (kN)	M_{max}(kNm)	σ(MPa)	τ (MPa)
Beam	17x14	596	21,07	9,85	- 1,75	3,21	0,62
Plumb	14x10	627	-70,24	-0,41	- 0,24	- 5,75	0,05
«*Travessanho*»	10x10	675	13,69	0,08	- 0,04	1,61	0,01
Diagonal	10X8	821	- 69,01	0,29	0,10	- 9,38	0,05

Table 3. Reductions of the tensions in wood bars

Elements	$(1 - \sigma_2 / \sigma_1) \times 100$	$(1 - \tau_2 / \tau_1) \times 100$
Beam	42%	14%
Plumb	42%	29%
«*Travessanho*»	68%	67%
Diagonal	34%	86%

4. Conclusion

The values obtained in the calculations confirm the efficiency of the Pombalino frontal design, since the obtained tensile values, although high, are below the rupture values of the wood material, confirming the high knowledge acquired by the architects of the Pombalino, although they did not have any theory of calculation and sizing, only had the observation of the collapsed buildings and the experimentation.

Through the linear elastic analysis of the "frontal panel", when subjected to the seismic action, characterized by EC8, it is possible to observe the exceptional relevance that the filling of the panels

of the wood structure with masonry assumes in the reduction of stresses and maximum displacements in the constituent elements and in the overall structural system. These advantages are also evident even when considering a reduced stiffness of masonry elements to take into account the reduction of the resistance of this material when subjected to seismic actions. In the present study, the value of 0.25 is adopted. According to various codes, stiffness modifiers used to simulate the behavior of structure members in cracked stage should range between 0.25 and 0.30. Considering the infill, with regard to maximum lateral displacements at the top of the frontal wall, the elastic analysis shows a reduction of Ux and Uy in the order of 1/3 and 1/2 respectively.
In the case of the bars with bigger efforts, it can be verified that the inclusion of mortar infill in frontal wall ensures significant reductions in the tensions of the constituent elements of the wood structural elements, as shown in table 3.
Equally and no less important is the conferred protection obtained by mortar infill with regard to fire resistance. In this context it is important to take into account that the construction system under analysis was developed to respond to the destruction that occurred in the city of Lisbon as a result of the earthquake of 1755 and the subsequent fires that broke out in the meantime.

5. Referencies

[1] Morais, António 2006 *Sistema Pombalino: Caracterização Estrutural e Comportamento Sísmico*. Revista Artitextos 03, Faculdade de Arquitetura de Lisboa
[2] Fialho, João Carmo; Morais, António José; Santos, João Lam; Svetlana Morais 2018 *Reuse in recovery of old buildings of brick and concrete debris as cladding mortar aggregates*, XVIII International Conference «Chemistry and Engineering Ecology» – at the Department of General Chemistry and Ecology of the Kazan National Research Technical University. KAI, Kazan from September 26th – 28th 2018;
[3] Morais, António José; Fialho, João Carmo; Santos, Lam dos 2019 *The 'pombalino frontal' design and its structural efficiency"*, CIVIL COMP 2019 Conferences, 16-19 September, Lake Garda, Italy;
[4] França, José Augusto 1977 *Lisboa Pombalina e o Iluminismo* (2.ª edição). Livraria Bertrand, Lisboa
[5] Euro código 8 *Projeto de estruturas para resistência aos sismos. Parte 1: Regras gerais, ações sísmicas e regras para edifícios*. NP EN 1998-1: 2010
[6] Cóias e Silva, V.; Alvarez, M. L.; Gomes, J.V.; Domingues, F. 1994 *Apoio à Modelação Estrutural por Métodos não Destrutivos de Paredes de Travamento dos Edifícios em "Gaiola" da Baixa Pombalina de Lisboa*. Conferência "2º ENCORE", LNEC, Lisboa.

Acknowledgments
This work is supported by CIAUD Investigation Center Of Architecture Urbanism And Design and by FCT. Foundation For Science And Technology.

IOP Conf. Series: Materials Science and Engineering 1148 (2021) 012008 doi:10.1088/1757-899X/1148/1/012008

The 'Pombalino Frontal' design and its structural efficiency

António Morais[a], Carmo Fialho, Lam dos Santos
The Lisbon School of Architecture, University of Lisbon, Portugal
[a]Abdiogo@hotmail.com

Abstract. After November 1st, 1755, earthquake which hit and destroyed Lisbon, the Portuguese military architects innovated the constructive techniques and conceived a 'new construction system' with anti-seismic characteristics, the so called '*Pombalino building*', often improperly named '*pombalino cage*'. Inserting in the '*Pombalino cage*' a vertical 'frontal' element, made up of Santo André Crosse's lattices, impresses the intellectual boldness, at that time, demonstrated by the military architects of the Kingdom, concerning the effectiveness and efficiency of the structural performance of the 'Pombalino building'. It should be noted that we were in 1755, so the concept of stress and deformation had not yet appeared, and, therefore, mankind did not yet discover the structural calculus theory, so the concepts of bending moment and normal effort were not known at that time. Although they did not know it, the frontal element design suggests their engineers' designers had an accurate knowledge about the phenomenology involved in their behavior, and, above all, they knew the optimal geometry to achieve structural efficiency to resist rupture of the 'Pombalino frontal' when an earthquake occurs.In this article we present structural calculations performed on a 'frontal pombalino' standard, used in the 'Pombalino cage', which demonstrate that the military architects, who developed the Pombalino system, had not only a perfect understanding of the St. André cross potential, but they also experimented with models to scale, because the 'pombalino frontal' geometric dimension is optimized in the plane of structural efficiency.

1. Introduction

The 'Pombalino system' was developed in the 18th century by the Portuguese Military Engineers as a structural response to the disastrous consequences of the great earthquake of 1755, given the widespread collapse suffered by most of the buildings in Lisbon, which caused thousands of casualties/deaths. The prime minister at that time was the Marquis de Pombal, a very enlightened and cult man, who took on the task of handling the reconstruction of the city of Lisbon after the earthquake, as the terrified king ran away from the capital, taking refuge in a countryside tent, named '*royal barrack of Ajuda*'.
The Marquis of Pombal, by his name *Sebastião José de Carvalho*, was a pragmatic and rational personality, little given to esoteric and religious explanations to justify natural disasters. The Marquis pronounced a prophetic phrase to the kingdom's military architects: "*the consequences of the earthquake are not a divine punishment; it is necessary to find a constructive solution to stand a new earthquake action*". Immediately after the earthquake, he ordered the police to close the entire destroyed area, barring access to the public for two years, including homeowners, transforming the entire area into a laboratory, where architects of the kingdom could interpret, understand and elaborate theories justifying the causes of the collapses.
As a corollary of this attitude and command of the Marquis, the kingdom's military architects innovated and created a new constructive system with capacity for seismic resistance - the '*pombalino cage*'. It was the first man-made world-wide system to withstand the action of an earthquake [1].

Observations and interpretations*in loco*, given that the destroyed area was transformed into a real laboratory, carried out by the military architects, headed by the Kingdom's master engineer, Manuel da Maia, allowed the acquisition of knowledge about the reasons for the collapse of those buildings under an earthquake action - the *phenomenology of collapse*.
The kingdom's architects learned from the ruins; they produced science. This knowledge led to the creation of a strong and potent structural system in the area of seismic resistance. The military architects rehearsed and developed a set of constructive, formal and material dispositions, seeking the new design of the 'Pombalino building' to achieve a greater capacity seismic energy dissipation and a better resistance to the horizontal actions.
Nowadays, the academic community in Portugal knows, with a reasonable degree of reliability, the functioning and structural behavior of the 'Pombalino building' existing in downtown Lisbon, however, we still know little, at the historical level, what technical and scientific assumptions are rational reached and assumed by the military architects of the 'Marquis of Pombal' in the development of the 'pombaline cage'. Due to the cultural, historical and economic importance of '*BaixaPombalina*' (downtown Lisbon) for the city, and the country, and also for humanity, in the field of seismic science, it is necessary to understand the evolution of the conceptual creative process that led to the definition and design of the pombalino cage constructive system.
In this paper we present the hypotheses and conclusions achieved through structural analysis and calculations based on the mechanical behavior of a standard frontal of the 'pombalino cage', which allow us to evaluate those military engineers' degree of comprehension, led by military engineer Manuel da Maia, on the effectiveness of the performance of the frontal present in the Pombalino building.

2. The Pombaline System

2.1. Constitution of the 'Pombalino System'

At the formal level the main characteristic of the Pombalino buildings is their regular geometric in plan and height, where it follows the building's symmetrical axis, respecting the two main characteristic axes in plan.
In the constructive dispositions they emphasize, among others, essentially two characteristic elements:
i) The '*Pombalino frontal*', consisting of a vertical truss of Santo Andre crosses, filled and covered with a mortar impregnated with pieces of ceramic material (clay mortar), connecting the facade walls in masonry, parallel to each other;
(ii) masonry exterior facade walls, with thickness on the order of the meter, containing a wooden structure inside (in the masonry interior), and reinforced on the corners of the building with high quality stone.
To such set (internal 'frontal' connecting parallel walls of facades in masonry) it is usually called '*pombalino cage*'. The designation of 'cage' comes from the three-dimensional arrangement of the wood elements, forming a space truss. This three-dimensional wooden structure, forming a cage, is considered by historians to be the great innovation of the 'Pombalino construction system'. Due to its great ductility, and given that the main task of a seismic structure is balancing and dissipating energy, transmitted by an earthquake, then, it is associated to the 'Pombalino frontal' the principal task of resisting an earthquake, and it is considered irrelevant and subordinate the presence of the outer walls in stone masonry. However, research shows that, in the overall functioning of the system, masonry walls' thickness plays a decisive role in the effectiveness of the system [2]. In fact, the power of the 'Pombalino system' results from the aggregate effect of the masonry walls, interconnected with each other, by the vertical wood frontals. The connection between the masonry walls, provided by the frontal walls, is vital for the static balance and entire building endurance. The military architects of the 'Marquis of Pombal' even carried out a seismic test of a model of the 'Pombalino building' on a natural scale in the '*Largo do Paço*' (Downtown Lisbon) [3].
The existence of vertical walls made of solid wooden truss inside in the interior of the building–the frontal, in addition to simple partition walls, without any special structural function, arranged in the two orthogonal directions of the rectangular plan of the building, carefully connected to the external

walls, the main masonry exteriors, the master-walls, and the floors, clearly demonstrates the intention of equipping the Pombalino building with capacity to support horizontal seismic forces.

2.2. Performance and behavior of a masonry

The 'Pombalino system' uses the materialization of the exterior walls, the facades, in a masonry constructive solution. It is, however, a material with a brittle and non-ductile behavior, therefore, with very low capacity to deal with tensile stresses, so it has a poor dynamic behavior, very susceptible to an earthquake action. It is known that in the case of structural masonry, as in the case of 'pombalino system façade walls', the resistance, to be guaranteed, essentially consists of the capacity to provide compressive strength and shear strength to the wall element, since these are the internal efforts to which the masonry is apt to work, given the poor ability to deal with traction by the masonry (traction is a great obstacle to masonry as a structural material).

The resistance to the seismic action of masonry buildings then comes essentially from the thickness of the constructive shape itself. Thus, massive walls acquire resistance to the traction provided by their own weight, as this cancels the vertical traction caused by a horizontal force (this is an artificial way of pre-stressing the wall).

In structural masonry buildings, the sturdy walls are implemented in the two main directions of the building plan, forming a three-dimensional structural system and developing a structural effect called by 'box effect'. In order to ensure the stability of the masonry building, it is important that the façade parallel walls of the building are strongly interconnected, in order to perform synchronous movements, ie, the oscillatory movement away from the plane in a given direction of a given wall, must be accompanied by equal movement, value and direction, executed by the other parallel opposite facade wall - the oscillatory movement must be absolutely equal in both walls, parallel to each other [2].

Forcing the facade walls parallel to each other to perform equal movements in value and meaning, was one of the discoveries that Marquis of Pombal's architects discovered in their on-site observation campaigns in the real laboratory of the 'Downtown of Lisbon' collapsed at the earthquake of 1755. They invented the wooden frontal, with crosses of Saint Andrew, to obtain this effect of equality in the oscillatory movement, between the facade walls parallel to each other. In the '*Pombalino building*', this coordinated and equal interaction, between the facade walls parallel to each other, is ensured by the vertical wood truss joining 'face to face' parallel walls - '*the frontal*'. The investigation carried out by A. Morais showed the determinant and specific role of the '*Pombalino frontal*' in the seismic resistance of the '*Pombalino* building'[2].

The presence of the frontal, by joining the outer walls of the two parallel facades, forces the oscillatory movement of these walls to move out of their plane, locking them mutually, consequently resulting not only in less displacement of the blocks of masonry, especially at the top of the walls, after the traction breaks in the joints, but forcing equal movement, so the internal pavements don't collapse. With this mechanism, the 'box effect' necessary to the overall operation of a masonry building is, although partially, developed and strengthened.

From the referred study [2], it is observed the '*Pombalino frontal*' presence interferes to the dynamic building behavior, however it is acknowledged that this influence does not have the relevance sometimes attributed, although its enormous deformation capacity, meaning a high ductile behavior. It is concluded from this study, that the 'Pombalino frontal' is not the preponderant element to the seismic resistance of the building availability; that is essentially assured by the masonry external wall façade's thickness, i.e., using the dead load as a resistance method - *a self-weighting mode of resistance*. The Pombalino frontal when forcing the facade walls to move together, prevents the floors of the building from falling off the walls and collapsing themselves. This is the principal role played by the frontal in the Pombalino building.

3. Structuralstudyofthe designofthe pombalino frontal

3.1. Frontal design

In historical terms, the design of the Pombalino frontal closes some doubts about the real knowledge held by the military architects about the effective functioning of the frontal and, especially, its formal and geometric design (crosses of Santo André and thickness of the wooden elements) [4]. At that time,

there was no structural calculation, nor the concept of stress had been created by humanity. It is surprising that, in the absence of the theoretical tools of structural calculation, the architects of *Pombalino* could invented a structure and a structural system with anti-seismic characteristics, where the frontal is presented with an optimized design for the performance of the seismic function in the *Pombalino*building.
This research intends to contribute to clarify the extent to which the architects of the Pombalino were aware of the efficient and effective frontal design they invented. Three models were then conceived for structural analysis, with of three different geometric designs, which was performed using the SAP 2000 program. The first model does not contain crosses from Santo André; only has vertical and horizontal elements. The other two models are identical in the formal plane, showing as difference the spacing between the vertical elements, which in the standard case of the Pombalino building is 1.0 meter while, in this study, the third model presents a spacing of 1.5 meters.

3.2. Characterization of seismic action in the frontal

The research methodology adopted to analyze the effects of the seismic action on the frontal consisted of an equivalent static analysis. For the scope and objective of the investigation, this approximation is reasonable because it is a regular structure in plan and height, with predictable dynamic behavior. Under these conditions, and because the EC8 [5] allowed it, the effects of the seismic action were determined by applying to the structure a set of horizontal forcesF_i,acting in parallel with the seismic action adopted and in the center of the respective masses, which in this study were concentrated at the level of the respective floors.
According to EC8 5.3.2, part 1.1, its values are given by the expression$F_i = \varepsilon_0 w_i$.
Where:
S_d (T_1) - ordered from the calculation spectrum for period T1;
T_1 - period of fundamental vibration for the movement considered;
m - mass of the building, corresponding to the area of influence of the front;
λ - correction factor ($\lambda = 0.85$).
In order to characterize the seismic action, the maximum acceleration of the terrain ($a_g = \gamma_I a_{gR}$) and taking into account the values for the maximum reference acceleration in the various seismic zones in mainland Portugal, if there is a Type 2 seismic action in a 2.3 seismic zone (Lower Pombalino), we will have $a_{gR} = 1.7$ ms^{-2}. It was considered the importance factor corresponding to current buildings, importance II $\rightarrow \gamma_I = 1$. Thus, $a_g = 1.7$ ms^{-2}.
The calculation spectrum S_d (T_1) for the study conditions, considering the values of $T_B = 0.10$ and $T_C = 0.3$ obtained in EC8 (Table 3.3) $T_B \leq T \leq T_C$, is defined by the following expression:

$$S_d(T) = a_g.S.\frac{2,5}{q}.\left[\frac{T_C}{T}\right] \quad \text{(EC8 3.2.2.5)}$$

Where:
a_g - calculation value of the acceleration at the ground surface ($a_g = 1,7$ ms^{-2});
S - soil coefficient (S = 1.8);
q - coefficient of behavior (q = 2,5) EC8 (table 8.2);
T_C - upper limit of the period at constant spectral acceleration plateau;
T - vibration period of a linear system with a degree of freedom

$$T = C_t \ .H^{\frac{3}{4}} \quad \text{(EC8 4.3.3.2.2)}$$

C_t - 0.050 according to EC8 4.3.3.2.2 (3);
H - front height.

3.3. 'Frontal Panel' geometric characterization

Since there were no elements to accurately characterize the geometry and constructive details of the original designs of the 'frontal wall', the structure presented in a paper at the conference "2nd ENCORE", in Lisbon 1994, (*Coias and Silva, Lourdes Alvarez, Víctor Gomes, Fernando Domingues*) [6] as a geometric proposal for the frontal structure (Figure 1) [6]. For the overall dimensions of the

'frontal wall', a framework with 12 x 6 m2 was admitted, characterizing a building of 4 floors including the ground floor.
The first model reproduces a frontal wall without crosses of Saint Andrew (Figure 2). The other two models are identical in the formal geometric plane, adopting both locking of the panels with crosses of Santo *André*. The difference between the latter two models lies in the dimension of the orthogonal mesh of the lattice structures, consisting of diagonal locking posts and crossbars in each of the panels. In one model, spacing of 1.5 meters between uprights and crossbeams were adopted (Figure 3); the sections of all bars being enlarged to obtain a total mass identical to that of the standard structure of the 'Pombalino building'. In the other model, spacing of 1.0 meter were used between uprights and crosspieces (Figure 4). This last model corresponds to the standard structure of the 'Pombalino building', according to the ones made in several works of '*BaixaPombalina*'.
The results obtained in these three models show the level of knowledge acquired by the architects of the Pombalino, since, as we shall see, the standard solution adopted in the Pombalino building is the optimized one in the seismic resistance chapter. It is surprising how these military architects, without theories of calculation, managed to reach this level of scientific knowledge, only using experimentalism observation.

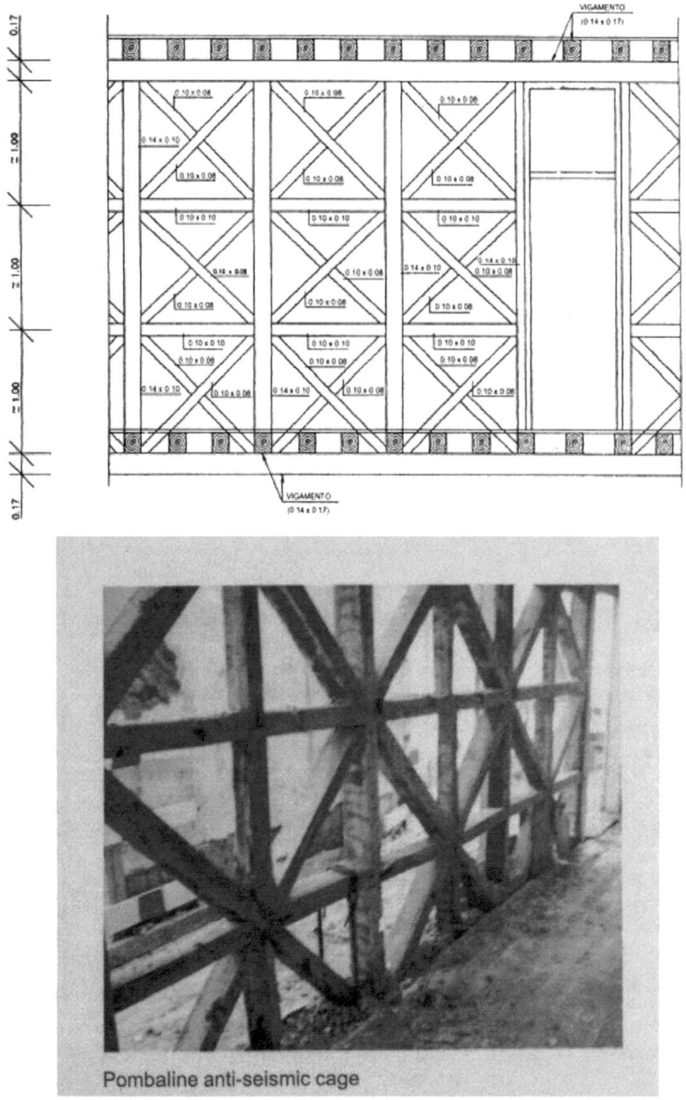

Figure 1. Geometry of the frontal structure

3.4. Frontal inertia forces

For the determination of the inertial forces acting on the frontal wall and for the EC8 [5], the model that substitutes the deformed of the fundamental mode by a linear configuration, considering the equivalent forces, at the level of the floor girder (Fi) resulting from the weighting of the basal force (F_b) by the parameter hi x mi, which results respectively from the height and mass corresponding to the level of the force i, obtaining for each of the floors the equivalent force given by the following expression:

$$F_i = F_b \times \frac{h_i \times m_i}{\sum_1^n h_i \times m_i} \quad \text{(EC8 4.3.3.2.3)}$$

c/ $F_b = S_d(T_1) m \lambda$

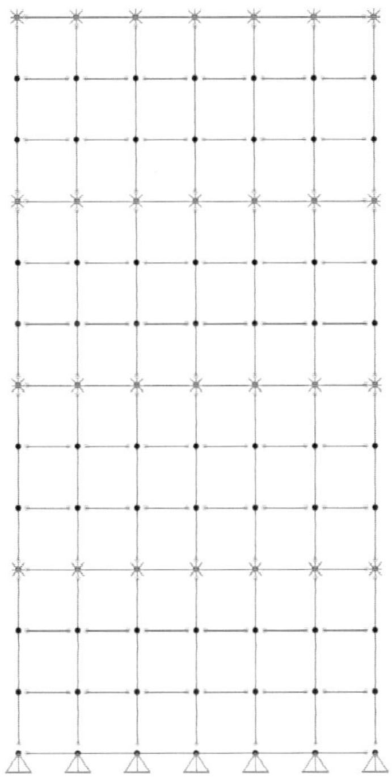

Figure 2. Model of frontal wall without crosses of Saint Andrew

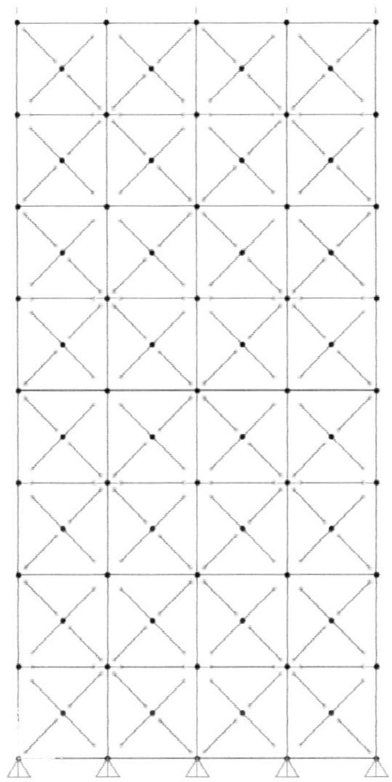

Figure 3. Model of frontal wall, with crosses of Saint Andrew, spacing of 1.5 meters between uprights and crossbeams

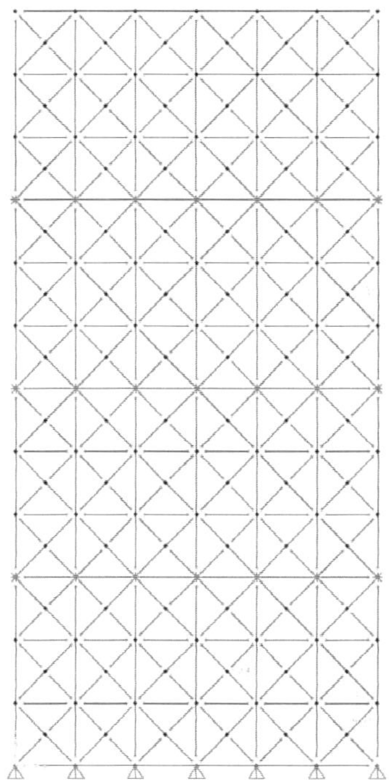

Figure 4. Model of frontal wall, with crosses of Saint Andrew, spacing of 1.0 meter between uprights and crossbeams

In the determination of the shear force at the base, we considered the masses of the frontal wall, of the houses and overloads, corresponding to the area of influence of the frontal. In this analysis, the mass values of the elements were obtained from the usual actions for dimensioning structures, according to the following assumptions:

a) Frontal wall (c = 6m; h = 3m)

It was considered that the frontal wall structures in analysis consisted of profiles of pine wood totaling 1 m^3 per floor.

$$W_{pf\,G} = 1\,m^3 \times 6\,kNm^{-3} = 6kN$$

b) Flooring

Assuming an area of influence of 6 x 5m^2 and considering the own weight of the wood structure 0,5 kNm^{-2} and the overload 3 kNm^{-2} we will have that, at the level of each floor, the force resulting from the weight of the floor and its overload will be:

$$W_{pav\,G+Q} = 6 \times 5 \times (0{,}5 + 3) = 105\ kN$$

From chapter 3.2 we obtain spectrum calculus$S_d(T)$

$$S_d(T) = a_g.S.\frac{2{,}5}{q}.\left[\frac{T_C}{T}\right] = 1{,}7 \times 1{,}8 \times \frac{2{,}5}{2{,}5} \times \left[\frac{0{,}3}{0{,}32}\right] \approx 2{,}8$$

Considering the previous components, it results that the cutting force in the base, according to the gravitational forces considered, will be:

$$F_b = S_d(T_1)m\,\lambda = 2{,}8 \times \frac{F_g}{g} \times 0{,}85 = 0{,}24F_g = 0{,}24 \times 444 = 106{,}56\ \mathrm{kN}$$

$$F_i = F_b \times \frac{h_i \times m_i}{\sum_1^n h_i \times m_i}$$

The forces of inertia Fi in the different floors are indicated in the following table (Table 1)

Table 1. Calculation of forces of inertia Fi in the different floors

Floor	h_i(m)	F_{gi} (kN)	F_{gi}x h_i (kNm)	F_i (kN)
4	12	111	1332	42,62
3	9	111	999	31,97
2	6	111	666	21,31
1	3	111	333	10,66
R/C	0			
Σ		444	3330	106,56

3.5. Characteristics of materials

We consider for wood (forces considered // to fibers)the acceptable values: $\sigma_{CMax} = 33MPa$; $\sigma_{TMax} = 33MPa$;

$v = 0{,}4$;$\rho = 0{,}6\ kgm^{-3}$;$E = 10GPa$;$\alpha = 5 \times 10^{-6}$

4.Results

4.1. Horizontal and vertical displacement at the top of the frontal

Based on the results obtained in the structural analysis of the three models considered in this investigation, described above, the displacements, the maximum stresses and the tensions in the wooden frontal bars were calculated. The calculated maximum lateral and vertical elastic displacements are observed at the ends of the frontal end, as expected:

• Frontal wall, with section 12 x 6 m^2, diagonals and reticulated mesh 1.0 x 1.0 m^2;

U_x = 0.232 m; U_z = - 0.008 m

• Frontal wall, with section 12 x 6 m^2 with diagonals and reticulated mesh of 1,5 x 1,5 m^2;

U_x = 0.012 m; U_z = - 0.006 m

• Frontal wall, with section 12 x 6 m^2 with diagonals and reticulated mesh of 1.0 x 1.0 m^2;

U_x = 0.009 m; U_z = -0.006 m.

From the results it can be seen that the military architects of the 'Pombalino' had knowledge about the effectiveness of the insertion of the crosses of Santo André inside the frontal, between the crosspieces

and the props, although at that time humanity still did not develop any rational theory of calculation and design of structures (not even the concept of state of stress and deformation had been invented). Note that not only the maximum horizontal displacement decreases from 23.2 mm to 9 mm, when the St. Andreas crosses are introduced, as also the vertical displacement decreases from 8 mm to 6 mm.
With a horizontal displacement of only 9 mm, we realized that the military architects of Pombalino achieved a remarkable achievement for that time, in the 18th century, as they succeeded in achieving the desired effect of equalizing the oscillating movements of the façade, walls parallel to each other, with a introduction of vertical trellis incorporating the crosses of Saint Andrew.
But they reached another level of knowledge; with a spacing of 1,0 meters the horizontal displacement is still significantly lower, since it goes from 1.2 cm to 0.9 cm. That is, they optimized the trellis design for the desired effect. This optimization effect is again evident in the calculation of the maximum stresses, as shown in the following paragraph.
The results obtained confirm the assertion expressed in the initial summary of this article. The military architects of thePombalino evolved in their knowledge, as they watched the collapsed buildings in the Downtown Lisbon. From this visual observation, and from the consequent reflection on the real causes of the collapse, they fabricated science and acquired knowledge about the real functioning of buildings during an earthquake. It was not a coincidence the innovation of the vertical frontal connecting the facade walls parallel to each other. They knew the importance of setting the parallel walls of the façade to oscillate in synchronous motion - it was necessary to avoid collapsing the interior floors. They achieved this with the insertion of the crosses of Saint Andrew inside the trellis. The front not only ensures the movement of both walls together, but also drastically reduces the horizontal displacement, forcing a drastic reduction of stress on the facade walls in masonry.

4.2. Efforts and tensions on floor 1 of the building

The tensile stresses were calculated only for vertical truss models with spacing of 1.0 m and 1.5 m, between uprights and crosspieces. In summary, the values obtained for the stresses and tensions in the pieces that present themselves with higher tensions are presented (Table 2), which are, as expected, at the level of floor 1.

Table 2. Maximum Loads in Frontal Bars

Element	Section	Dim. (cm2)	Bar	N (kN)	V_2 (kN)	M_3(kNm)	σ(MPa)	τ (MPa)
Beam	1X1	17x14	596	64,3	11,7	- 2,3	6,1	0,7
	1,5x1,5	20x20	68	44,3	16,2	- 4,2	4,3	0,6
Plumb	1X1	14x10	621	- 95,4	0,01	- 0,01	6,9	$1,5x10^{-3}$
	1,5x1,5	17x10	106	-125,5	≈ 0	- 1,05	9,6	≈ 0
«Travessanho»	1X1	10x10	663	53,3	≈ 0	0,02	5,4	≈ 0
	1,5x1,5	14x10	81	27,4	0,2	0,06	2,1	$2x10^{-2}$
Diagonal	1X1	10X8	729	- 45,4	0,2	0,02	5,8	$3,1x10^{-2}$
	1,5x1,5	10x10	129	- 43,6	0,1	0,02	4,4	$1x10^{-2}$

The results confirm the high structural knowledge in the field of seismic resistance achieved by the architects of Pombalino. We know from the various historical records that, in order to optimize the operation, have carried out tests on physical models in the search for the efficient and efficient design of the frontal. Only the experiment allowed them to reach the standard 1.0 m spacing at the frontal. As shown in the above calculations, with the standard spacing of 1.0 m, the maximum tension in the plumb bob is 6.9 MPa, whereas in the case of the 1.5 m spacing it is 9.6 MPa.
The values obtained in the calculations confirm the efficiency of the Pombalino frontal design, since the obtained tensile values, although high, are below the rupture valuesof the wood material, confirming the high knowledge acquired by the architects of the Pombalino, although they did not have any theory of calculation and sizing, only had the observation of the collapsed buildings and the experimentation.

5. Conclusion

Considering the results obtained, it is concluded that those military engineers who made the *'Pombalino cage frontal'* design, had a clear understanding of the overall structural and seismic behavior of the building, especially the need to interconnect and limit the displacements of the exterior walls. Although, they did not have theories of calculation and sizing, those military engineers of the *'Pombalino'* design realized the necessity of introducing crosses of Santo André inside the vertical lattice of the frontal and managed to optimize the geometry and sections of this vertical trellis.
The *´Pombalino Building'* innovation was a relevant intellectual leap. It was the first anti-seismic system engineered by mankind and, surprisingly, even today it follows the national seismic regulation now in force in Portugal.

6.Referencies

[1]Tobriner, Stephen (2004) – A GaiolaPombalina. O sistema de ConstruçãoAntissísmicoMaisAvançado do Século XVIII. RevistaMonumentos n.º 21, Direcção-Geral dos Monumentos e EdifíciosNacionais, Lisboa.

[2] Morais, António (2006) – SistemaPombalino: CaracterizaçãoEstrutural e ComportamentoSísmico. RevistaArtitextos 03, Faculdade de Arquitetura de Lisboa.

[3] França, José Augusto (1977) – LisboaPombalina e o Iluminismo (2.ª edição). Livraria Bertrand, Lisboa.

[4] Ferrão, Leonor (2017) – Eugénio dos Santos (1711-1760), arquiteto e engenheiromilitar: Cultura e prática de arquiteturaem Portugal no século XVIII; by the book.

[5] Euro código 8 – Projeto de estruturasspararesistênciaaossismos. Parte 1: Regrasgerais, açõessísmicas e regrasparaedifícios. NP EN 1998-1: 2010.

[6] Cóias e Silva, V.; Alvarez, M. L.; Gomes, J.V.; Domingues, F. (1994) – Apoio à ModelaçãoEstruturalporMétodosnãoDestrutivos de Paredes de Travamento dos Edifíciosem "Gaiola" da BaixaPombalina de Lisboa. Conferência "2º ENCORE", LNEC, Lisboa.

IOP Conf. Series: Materials Science and Engineering 1148 (2021) 012009 doi:10.1088/1757-899X/1148/1/012009

Ecological planning of the urban residential area for the sustainable development

Ning Kang[1, a], Yan Liu[1,b] and Zhihong Wang[1, c]
[1]School of Art, Hubei Polytechnic University, Huangshi, 435003, China
[a] kangning@hbpu.edu.cn, [b] 121100926@qq.com, [c] 254868344@qq.com

Abstract. Residential area is closely related to human living environment, and the construction of ecological residential area has become an important part of current architectural research. Studies on the landscape planning and design development direction of urban ecological residential area from the perspective of sustainable development were made in this paper, and then taking the planning and design project of "Cihu Lake Peninsula of Country Garden " in Huangshi city of China as an example, discussions about how to adjust measures to local conditions, put people first and keep pace with the times in landscape construction of eco-residential area were also present. This study provides reference for the combination of theory and practice of eco-residential planning and design.

1. Introduction

With the increasing environmental problems in the world, the urban ecosystem is increasingly fragile, and the residential environment is also facing many problems [1]. In recent decades, researchers at home and abroad have begun to pay attention to improving the living environment and realizing sustainable development. Therefore, ecological residential area has become the primary goal of urban residential area construction at home and abroad, and ecological residential area planning has also become a hot topic of research.

In the residential area planning design, instead of abandoning the artificial environment and natural environment antagonistic relationship, it should make the living environment integrated harmoniously with the natural environment, which is guided by the natural integrity and continuity of natural process. The urban eco-type residential area adopting this design concept is an ideal urban residential area with highly unified and sustainable development of natural ecology and human ecology, natural environment and artificial environment, material civilization and spiritual civilization.

2. The connotation of urban ecological residential area

2.1. Adjusting measures to local conditions

The respect and exploration of the local natural conditions, such as the climate and ecology, the historical and cultural conditions, are not subjective creations of the designers, but a comprehensive analysis of the living function and rules of the residential areas [2]. Only through systematic study of natural and human conditions, scientific grasp of modern production technology and comprehensive consideration as well as based on local specific conditions, respect the regional climate and the customs and culture, can we integrate the ecological concept into the details of planning and design and create a high-quality residential area with meticulous creation. Therefore, in the design of ecological residential areas, the primary task is to analyze the local natural factors, functional factors, ecological factors, human factors, spatial factors, technical factors, etc., and to make ecological planning and design suitable for the region in consideration of its unity, integrity and continuity.

IOP Conf. Series: Materials Science and Engineering 1148 (2021) 012009 doi:10.1088/1757-899X/1148/1/012009

2.2. People oriented

"People-oriented" is not a slogan of designers, nor a gimmick of developers. To truly achieve "people-oriented" firstly is to ensure the safety and health of residents, including traffic safety, personal safety and public health. In addition, to pay attention to environmental conditions, including clean and quiet environment, fresh air and clear water. Moreover, to ensure the quality of social interaction including harmonious, fair and mutual respect among neighbors, and pride in residential areas. And to ensure the enjoyment of leisure and entertainment, aesthetics and existing unique cultural and environmental resources, including the surrounding natural environment and cultural environment.

The ecological residential area is not just a visual form on the surface, but a material and emotional exchange generated from the harmonious relationship between human and architecture. Therefore, in the planning and design of ecological residential areas, people orientation is mainly reflected in residents' material and spiritual enjoyment, security and convenience of living, and the safety, convenience and comfort of their residences are guaranteed to the greatest extent. Considering the physiological and psychological needs of different groups of people, designer should enrich the functions and ornamentation of leisure facilities and service facilities, so that the old, middle-aged and young people will all feel physically and mentally pleasant. In addition, designers should also pay attention to the harmony and unity of the environment and the surrounding environment, such as the size, scale, details, texture, color and other forms of beauty.

2.3. Keeping pace with the times

Keeping pace with the times in the planning and design of ecological residential areas is to adopt appropriate new materials and technologies to achieve sustainable development. First of all, the whole life cycle of the architecture, from the initial planning and design to the subsequent construction, operation and management and final demolition, should be paid enough attention. The relationship between the building materials and the surrounding environment should be fully considered, and detailed layout should be made in the design stage to ensure that the materials and processes do not bring loads to the environment. Secondly, learn from advanced experience at home and abroad, optimize the allocation of water circulation and air circulation in the region, and maximize the saving and use of local materials and resources. Thirdly, intelligent technology and intelligent system should be reasonably adopted to improve the safety, comfort and convenience of residential areas. Therefore, keeping pace with the times in the planning and design of ecological residential areas is embodied in mastering relevant technologies and cutting-edge design, making efficient and rational use of resources, saving energy, land, water and talent, minimizing the impact on the environment, and providing high-quality living, working and activity space for residents.

3. Practice in ecological planning of urban residential area

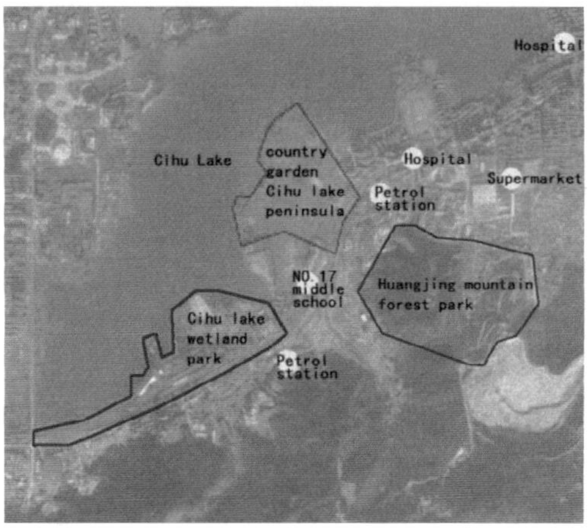

Fig 1. Aerial view of Cihu Lake Peninsula of Country Garden

3.1. Project background and overview
The Cihu Lake Peninsula of Country Garden Project is located in Huangshi City, which is close to a rare inner-city ecological lake named Cihu in China. The whole project is located on the south lakeshore of Cihu lake, adjacent to the Cihu wetland park in the west as shown in Figure 1. The project covers an area of 22.9 hectares with a building area of about 600,000 square meters and a plot ratio of 2.62. The total number of semi-detached villas, small and high-rise villas is 3400, with 3 floors, 10 floors and 11 floors respectively. There are no super high-rise buildings. Cihu Peninsula residential area is a top luxury residential project built by Country Garden on the bank of Cihu Lake in Huangshi, the vice central city of Wuhan. In addition to the strong brand guarantee of country garden, luxury community, old-age real estate, waterscape real estate and ecological livability are its project features. In the process of planning, humanized design and extreme construction technology were introduced into the community. This project is the closest to the lake in the whole Huangshi area and even in the whole country. The nearest place is only 55 meters, which is a real front line near the lake.

3.2. The ecological residential area planning of Huangshi Cihu Lake Peninsula of Country Garden
Cihu Lake Peninsula of Country Garden is located on the south lakeshore of the Cihu which is in the shape of peninsula with a plat terrain. And the view is excellent, because of three sides facing the lake. In the heart of the bustling city, a piece of heaven and earth is opened up where the original ecological lake is embedded in. It nearly takes 1 kilometers of natural ecological landscape along the lakeshore with watery blue reaching far beyond the horizon. In addition, the location of Peninsula is adjacent to the Cihu lake wetland park where it can be said that it looks like the new lung of the city, breathing with the city. Moreover, the surrounding public service facilities are relatively complete, such as education, medical care, commercial services, financial posts, telecommunications, municipal public administration, cultural and sports services and other supporting services are all within a service radius of 1km, which is the best combination of convenience, comfort and tranquility. The house types include three bedroom A109 square meters, three bedroom B123 square meters, three bedroom C125 square meters, four bedroom A142 square meters, four bedroom b143 square meters, four bedroom C189 square meters, five bedroom a262 square meters, five bedroom b273 square meters, five bedroom c278 square meters, villa a273 square meters, villa B278 square meters, villa c706 square meters, seven bedroom 706 square meters and other different types, which can meet the require of different families.

3.3. Low plot ratio and high afforesting rate
As shown in Figure 2, the buildings of Cihu Lake Peninsula of Country Garden are not more than 11 floors. The planning plot ratio of 2.62 and afforesting rate of super elevation 52% gets the whole city proudly. The villas with 3 floors are built up near the lake in the residential district planning and design which, on the one hand, can ensure the residents sufficient sunshine and good ventilation. On the other hand, can segregate the residents from the city's main road. This means that it protects the people away from noise, harmful gases and soot pollution. The project planning is mainly based on native tree species and adjusts measures according to local conditions to do their best to reserve the original vegetations. Planting trees depend on the seasons and are use of gently slope revetment afforestation and focus on the safety, functionality and aesthetics. The planning also adopts new technologies in order to give people a sense of comfort, including three-dimensional greening and roof greening to protect the privacy of residents, reduce the urban heat island effect and improve the quality of housing. Living in such a high greening rate of the community, owners can easily see the green scene, which will make their mood better. In addition, the places where a large number of plants are planted are usually the communication space and activity place for residents. Owners can walk, rest, play cards and chess in their spare time, which virtually enhances the feeling between neighbors. Green plants can also purify the air through photosynthesis, so that people who have been in the noisy city for a long time can enjoy a green and healthy life after work, which also makes the residential area more attractive.

Fig 2. Rendering of Cihu Lake Peninsula of Country Garden

3.4. Organic architecture and intelligent decoration

Organic architecture is a faction of modern architectural trend, represented by American architect F.L. Wright, who regarded architecture as an organic life and in the process of continuous development [3], [4]. Organic architecture is an emotion planted in the nature of life and natural forms. It can draw nutrition from the vitality of the natural world and its various biological forms and processes. Cihu Lake Peninsula of Country Garden adopted the concept of organic architecture to form a harmonious and symbiotic relationship between the building and the surrounding environment, and to meet the requirements of sustainable development in four ecological systems, namely, the systemic view of the integration of nature and man, the natural-view of "Nature, the principles of", the economic view of wonderful workmanship excelling nature and the humanistic view of human-oriented. Cihu Lake Peninsula of Country Garden was in combination with environmental terrain and user demand with fully considering the natural and human factors and the layout of the planning and architectural appearance towards family harmony with surrounding environment. Strictly control the space of high-rise buildings was to ensure sunshine time, and a large number of environmental protection materials, local materials and high cost-effective materials and intelligent decoration were adopted to truly achieve coordination and symbiosis between architecture and environment.

Moreover, intelligent decoration was widely used in residential planning and design. That is to say, integrate modern information technology into the home decoration project to make the family decoration and functional information combination [5]. The traditional water meter, electricity meter, gas meter and so on were independently controlled by the community center, so that people do not need to spend time for checking the meters or paying fees. All the necessary wires and networks were preinstalled in advance and it also has the function of real-time monitoring and alarming and reacting intelligently according to different alerts. The finished residence with intelligent decoration is helpful to users to reduce the pollution and disputes in the process of individual household decoration.

3.5. Rain and sewage diversion and low carbon emissions

Due to the light pollution of rainwater, it can be directly discharged into the urban river after diversion, and then it can be used as natural landscape water after natural precipitation, and it can also be used as urban municipal water for spraying roads. Therefore, the use efficiency of surface water can be improved when rainwater flows into rivers through purification and buffering. At the same time, the sewage is discharged into the sewage pipe network and treated by the sewage treatment plant to realize the regeneration and reuse of sewage. The biggest advantage of rain and sewage diversion is that it can speed up the sewage collection rate, improve the sewage treatment rate, avoid the pollution of sewage to river and groundwater, significantly improve the urban water environment and reduce the cost of sewage treatment. Another advantage of rainwater and sewage diversion is that it is convenient for rainwater collection and utilization, centralized management and discharge, and reduces the impact of water quantity on the sewage treatment plant, so as to ensure the efficiency of the sewage treatment plant. In this way, the application of rain and sewage diversion technology will greatly improve the environmental quality, grade and property management level of residential areas, and effectively improve the living environment and quality of life of owners.

The rain and sewage diversion system were planned and set up to collect the domestic sewage in the region through the sewage pipe network and transfer it to the sewage plant for centralized treatment and discharge after reaching the standard. In addition, rainwater was collected by rainwater pipe network and then recycled for green irrigation in the district. Besides this, low-carbon emission and environmental protection policies were carried out to the end in the planning by using central heating to reduce air pollution. The garages were equipped with charging piles for electric vehicles and new energy vehicles to facilitate green travel and separation of people and vehicles was made above and below ground so as to improve air quality. The diversion of rain and pollution and low carbon emission in the planning and design of ecological residential area on the Cihu Lake Peninsula of Country Garden had become a landmark demonstration project for the transformation of resource-exhausted city.

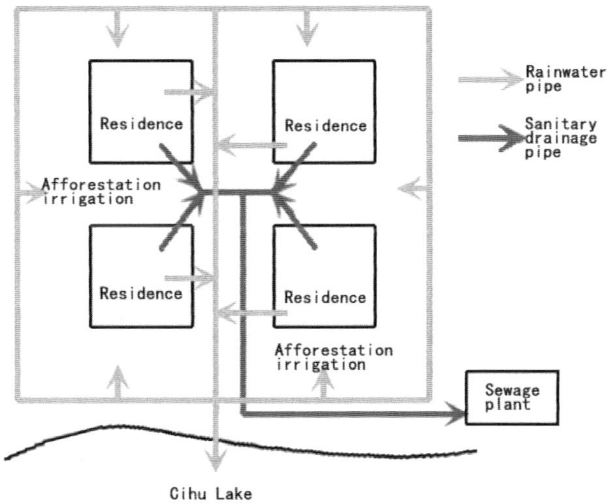

Fig 3. Schematic diagram of rain and sewage diversion

4. Results

Ecological design of the Cihu Lake Peninsula of Country Garden emphasizing the mutual fusion and mutual effects and a harmonious relationship between man and nature, clever solved the problems such as energy-saving, land-saving, water-saving and environmental protection. The main idea of the design was to make use of natural ecological process and law of recycling to achieve the sustainable development of human and nature harmonious coexistence, and eventually improve the quality of human living, working, leisure, learning and entertainment, which has important theoretical and practical significance for the sustainable development of residential areas.

5. Summary

As is well known, the carrying capacity of natural resources is limited. Therefore, planning for residential areas shall reasonably determine the population density of residential areas and the intensity of land use. The planning and design of ecological residential areas should make the best use of natural resources such as land, sunshine and water. People should spare no effort to conserve non-renewable resources and guarantee the regulation and self-purification of the natural ecological system and minimize the influence of living activities on the environment. It is in order that urban residents live more comfortably, conveniently and ecologically. Only in this way can we achieve the harmonious coexistence of man and the nature.

6. Acknowledgement

This work is funded by Opening fund of the research center for mining and metallurgy culture and economic and social development in the middle reaches of Yangtze river, Hubei Province

(No.2016kywh12), and The Youth Project of Philosophy and Social Sciences in Colleges and Universities of Hubei Province (No.20Q149). All the satellite images were provided by Google Earth.

7. References

[1] Ahern J. Novel urban ecosystems: concepts, definitions and a strategy to support urban sustainability and resilience [J]. Landscape Architecture Frontiers, 2016, 4(1): 10-22.

[2] Childers D, Cadenasso M, Grove J, et al. An ecology for cities: A transformational nexus of design and ecology to advance climate change resilience and urban sustainability [J]. Sustainability, 2015, 7(4): 3774-3791.

[3] Panagopoulos T, Duque J A G, Dan M B. Urban planning with respect to environmental quality and human well-being [J]. Environmental pollution, 2016, 208: 137-144.

[4] Chang J R. Defining a Novel Meaning of the New Organic Architecture [J]. A+ BE| Architecture and the Built Environment, 2018 (1): 143-172.

[5] De Jong, Martin, et al. "Sustainable–smart–resilient–low carbon–eco–knowledge cities; making sense of a multitude of concepts promoting sustainable urbanization." *Journal of Cleaner production* 109 (2015): 25-38.

AUTHOR INDEX